ALSO BY ERIC P. WIDMAIER

Why Geese Don't Get Obese (and We Do)

The Stuff
of Life

The Stuff of Life

Profiles of the Molecules That Make Us Tick

ERIC P. WIDMAIER

Illustrations by Heather Keller

TIMES BOOKS

Henry Holt and Company • New York

Times Books
Henry Holt and Company, LLC
Publishers since 1866
115 West 18th Street
New York, New York 10011

Library of Congress Cataloging-in-Publication Data
Widmaier, Eric P.
    The stuff of life : profiles of the molecules that made us tick / Eric P. Widmaier ;
illustrations by Heather Keller.—1st ed.
    p. ; cm.
    Includes index.
    ISBN 0-8050-7173-3 (hb.)
    1. Biochemistry—Popular works.   I. Title.

QP514.2 .W497 2002   2002019735
572—dc21

Henry Holt books are available for special
promotions and premiums.
For details contact:
Director, Special Markets.

First Edition 2002

Designed by Paula Russell Szafranski

Illustrations by Heather Keller

Printed in the United States of America

1   3   5   7   9   10   8   6   4   2

To Maria, every molecule of whom I love

CONTENTS

PREFACE

With the sequencing of the human genome in 2000, biologists estimate that there are about 30,000 to 40,000 different genes in the human genome, fewer than originally thought but still a daunting number. Although the sequence of the genome is now available, this is a far cry from understanding what the functions of all these genes are. Yet, it is a beginning.

A gene is a bit of DNA packaged within a chromosome in the nucleus of a cell. Most genes are active only at distinct periods during life. During those times, the genetic code within the DNA is "transcribed" into a signaling molecule called RNA. A colleague of mine, Dr. Hershel Raff, often says that transcription is like changing forms within the same language—like transcribing spoken English into written English.

In effect, the RNA is a mirror image of a gene's DNA. The RNA directs the formation of a protein, by a process called translation. To continue Dr. Raff's analogy, translation is also like changing languages—converting RNA to protein is akin to translating German into English. The protein that is formed is known as the product of the gene. And genes themselves have no activity or function other than to serve as a readily available source for making proteins. But that is certainly a worthwhile function, since the proteins are what govern the ability of a muscle cell to contract, a heart cell to beat, or a brain cell to store memory.

Given the enormity of the genome and its 30,000 or so gene products, choosing a manageable number of proteins and other molecules to highlight in this book was not easy. I have chosen to focus on those biologically important molecules whose chemistry and function are especially well understood and important, and on the relationship of these molecules to the human condition in both health and disease.

Life is not only protein, however. Our bodies depend on numerous carbohydrates, lipids, small cholesterol-based steroid hormones, vitamins, ions, gases, water, and neurotransmitters. I've selected a variety of these important molecules to round out the picture of how chemistry determines physiology, and how physiology determines health. We begin by examining the nature of the functional units of all biology, the atoms and molecules. This is followed by an exploration of the so-called blueprints of life, our genes, and the proteins that are formed from genes. How different proteins and other molecules work together to create a functional, healthy person makes up the middle of the book. We conclude by discussing two systems—the nervous and endocrine—that are the key regulators that keep all the other parts of the body working in harmony.

The Stuff
of Life

[1]

The Blueprints

The human body is built in layers of complexity. Viewed in its entirety, the body's different parts can be observed to work together in an integrative way. The brain controls the function of the heart, which in turn controls the function of the muscles and other organs. Likewise, the pituitary gland controls several hormone glands, which in turn determine electrolyte balance, blood pressure, and sugar metabolism. On a less global scale, each of the organs in the body (the liver, kidneys, skin, heart, and so on) are themselves composed of smaller units. Within the kidney, for example, specialized cells regulate the balance of salt and water in our bodies. Other groups of specialized kidney cells secrete hormones, or filter blood to remove waste products.

On an atomic scale, however, even a single cell is an entire universe. Within cells are smaller structures that make proteins, package chromosomes, and generate energy. To understand how these tasks are accomplished, we must look to the molecular level. A large protein molecule, for example, is composed of many smaller molecules called amino acids. But even an amino acid is not the smallest functional unit in a protein. Within any amino acid (or sugar molecule, or oxygen molecule, or any other molecule) are a number of atoms.

Just what is an atom, anyway? Every element, like hydrogen, nitrogen, and oxygen, is composed of atoms. Atoms are the smallest functional unit of matter. That is, an atom can be broken into subatomic (smaller than an atom) particles if sufficient (enormous) amounts of energy are provided, but those smaller bits of matter do not by themselves participate in biologically meaningful reactions. An atom can be crudely envisioned as a sort of mini-solar system, although in reality an atom's structure is much more complex and less orderly than a solar system. At the center of the system is the nucleus, composed of subatomic particles called protons and neutrons. Protons carry a positive electric charge, while neutrons have no charge (they are electrically neutral). Surrounding this densely packed nucleus are anywhere from 1 to more than 100 electrons, arranged in increasingly wide and complex orbits around the nucleus. The electrons are only a tiny fraction of the size of the nucleus and carry a negative electric charge.

Each atom has the same number of electrons as protons, which allows the two opposite electrical charges within the atom to cancel each other. Let's imagine a relatively small atom such as carbon, which contains a nucleus with six protons. It will, therefore, have six electrons orbiting around it. Although that gives an atom a neutral electrical charge, it turns out that there is "room" for two additional electrons to spin around a carbon nucleus. That's because electrons are arranged around the nucleus in poorly defined, mutually exclusive orbits, each of which has a predetermined capacity to permit extra electrons to buzz around within that orbital sphere. Having as close as possible to a full complement of electrons in the outermost shells of an atom increases a molecule's stability. So if two atoms—say, carbon and oxygen—come together under the right circumstances, they may "share" some of each other's outermost electrons. In that way, both atoms will appear to have filled up their empty "electron slots." This is true because the speed at which electrons zip around their orbital shells is so fantastic that it makes little difference if the shell is enlarged a bit by the merging of two or more atoms. When this happens, we say that the two atoms have formed a chemical bond and have joined to make a molecule. One atom of carbon and one of oxygen, by the way, would yield the poisonous molecule carbon monoxide.

It is common in nature to find two or more different kinds of atoms sharing electrons and combining to create a new, larger substance, a molecule. Some molecules are quite simple. Water, for example, is composed of one oxygen atom combined with two hydrogen atoms. Others are extremely complex; a protein may be composed of hundreds of amino acids, and each amino acid may be composed of several atoms of nitrogen, carbon, oxygen, hydrogen, and sulfur. Thus, molecules can be broken down into atoms, but atoms cannot be broken down into any other functional unit.

In the world of the molecule, anything heavier than about 0.0000000000000000000001 (one ten-billionth of one-trillionth) grams is considered large. If that number seems meaningless, then consider that there are about as many molecules of water in a drop of blood as stars in the known universe!

It may be a cliché, but proteins are truly the building blocks of all life. They are the cinder blocks and the 2x4's of our cells. While we hear about DNA, it only exists to direct the making of proteins. But proteins don't just provide the body with physical structure, they also catalyze chemical reactions, taxi gases like oxygen through the blood, and produce energy. Enzymes are also proteins. Enzymes are molecules with very precise three-dimensional structures, which allow them to interact with other molecules. In some cases, this interaction produces the destruction of another molecule. In other cases, enzymes help fuse two simple molecules to produce one complex molecule.

Different species of the same phylogenetic class (for example, mammals) share much of the same DNA, or at least DNA that is recognizably similar. And even species that on the surface bear little or no resemblance to each other have much of their DNA in common. The lowly nematode worm shares roughly 40 percent of its DNA sequence with that of humans. As you move up the scale of animal complexity, the similarity increases, of course, so that by the time you reach the other primates, such as chimps, the similarity to human DNA approaches 98 percent. Unrelated people share 99 percent similarity, and related people 99.5 percent. We are not as different from one another as we may think.

Despite this similarity, a relatively small amount of different DNA can make vast differences in the appearance and behavior of an organism. A single molecule of DNA may contain hundreds or thousands of separate functional units, called genes. Each gene is a strip of DNA distinguished from the next strip by telltale regions that signal the beginning of a new gene. The enzyme responsible for converting genes into RNA recognizes these starting positions. Every cell in our bodies has the exact same DNA, and thus the exact same set of genes. But cells in our skin, for example, have an active gene for a fibrous protein called keratin, which is the basis of skin. This same gene is inactive in most other cells, and this prevents keratin, and therefore skin, from showing up in, say, the liver or bone marrow. The ways in which a particular cell is able to activate only a specific subset of genes and not others is obviously of enormous importance to scientists, who are only beginning to find the answers. This question holds the key to understanding how an organism develops from an immature, undifferentiated embryo of just a few cells into a fully functional adult animal with, in the case of humans, trillions of cells. On a more practical level, it holds the key to regenerating lost or damaged tissues and having them function and appear like the original.

The discovery that the DNA molecule exists in the form of a twisted helix and contains only four major chemical elements, repeated in various arrays, was the landmark event that ushered in the field of science known today as molecular biology. That discovery has allowed us to begin understanding how genes can be active at one time and silent at others; how simple changes (mutations) in any of the four chemical elements of DNA can lead to the formation of abnormal proteins, or even premature death or failure of an embryo to develop; and how genes might someday be modified or even changed to correct human disease. All of the molecules discussed in this book are either formed from genes or act on genes to control their activity. Thus, it is fitting that we begin with a look at the chemistry and physiology of DNA, RNA, and the substances they make—proteins.

DNA and RNA

Chromosomes, genes, DNA, RNA—we hear these terms often, and for most of us they present something of a mystery. But DNA and RNA are just molecules made up of atoms linked one to the other, just like all molecules. Unlike many molecules, though, DNA is extremely long because of all the information it contains. In fact, to package all of our DNA into the nucleus, or command center, of a cell, the DNA must be folded, twisted, and folded again into a compact shape. Were it not, a single DNA molecule would be several inches long. Considering that the nucleus of one of our cells is only about 1/5,000th of an inch wide, it is easy to see why DNA must be so compact.

DNA (deoxyribonucleic acid) is made up of a molecule of sugar (ribose), some phosphates (phosphorus and oxygen bound together), and a group of four molecules known as bases. The latter are relatively simple molecules that are able to attach to one end of the ribose molecule. The phosphate groups also attach to the ribose, but at the opposite end. Phosphate groups are highly reactive and will link one ribose to another in a sort of linear chain. Thus, a molecule of DNA "grows" as one ribose—with its attached phosphates and bases—links up with another ribose, and so on.

To complete the molecule, each base forms weak electrical attractions to another base that is complementary in structure on another ribose-backed DNA chain. The two chains come together to form a sort of ladder-shaped molecule, with the bases forming the rungs of the ladder. As you walk along these rungs, you pass from one gene to another.

The linking together of two strands of DNA like this causes a physical strain on the molecule, and the double chain of DNA twists itself into a helical shape. This gives the molecule a certain stability, and the DNA can now be further contorted into smaller and smaller volumes. Thus, it wraps itself around proteins (called histones) found in the cell nucleus and continues folding and refolding on itself until all 100 million or so rungs of the ladder fit into the tiny cell nucleus. We call a single DNA molecule wrapped up in this configuration a chromosome. Different animals have different numbers of chromosomes in each cell; humans have a total of 46, all of which must be condensed in this way.

Why should a DNA molecule need to be so extraordinarily long? The function of DNA is to "code" or hold the blueprint for all of the body's proteins. But though there are upward of 3 billion base pairs (the rungs) along the chromosomes, only about 30,000 or so proteins are made from all that DNA. Clearly, the math does not add up.

As we've already learned, the coding within the DNA is packaged in units called genes. Each gene is a smaller bit of a chromosome, and the number of genes along a chromosome matches the number of proteins formed from that chromosome. Within a gene it takes three bases to code for one amino acid, the individual building blocks that make up all proteins. There are only four bases: guanine (G), cytosine (C), adenine (A), and thymine (T). Thus, a sequence of CTG along a gene codes for an amino acid called leucine, while a sequence of CGG codes for one called arginine. Whenever those sequences are present, a leucine or arginine molecule will be added to a growing protein. The next set of three bases will determine what the next amino acid will be in the protein, and so on. So part of the mystery of the "extra" DNA lies in the fact that you need three bases for each amino acid.

Oddly, there remains a very large amount of DNA that is not a code for anything. Some of these regions of DNA are known as introns, and they are interspersed within most genes. Although much of the DNA in a chromosome may be introns, the function and evolutionary significance of introns are still unknown.

Why did DNA need to have a second strand in the first place? This evolutionary milestone ensured that whenever a cell divided into two new cells, each would receive a full complement of the parent cell's DNA, with all of its genes intact. That's because when a cell divides, the two DNA strands split along the rungs of the ladder, and each new cell gets its own strand. By the time cell division is complete, the duplicate strand of DNA has been resynthesized in each new cell. The two new strands of DNA come together, the DNA folds up, and two new, fully functional cells are born. In other words, the second strand is what makes heredity possible.

One last thing of great importance is how the encoded bases in DNA get translated into amino acids. How does a cell "know" that CTG is the right DNA sequence for leucine? Two intermediates must assist in this translation from one type of molecule (DNA) into another (protein). When a certain protein, such as an enzyme, is needed, that protein's gene is activated. Typically, the cell cytoplasm senses the protein deficiency and shuttles a signaling molecule to the cell's nucleus. There the signal finds the correct gene and begins the process of "unwinding" the coiled DNA.

As the DNA unwinds, the gene becomes exposed. Enzymes in the nucleus split the two-stranded DNA along the rungs, and a mirror image of the gene is created using available bases, riboses, and phosphate groups. This mirror image differs a bit from DNA, because the ribose has an extra oxygen atom (which is why it's called ribonucleic acid, RNA, instead of deoxy-ribonucleic acid), and because it uses a base called uracil (U) instead of thymine. Other than that, it's essentially the same as, just shorter than, a molecule of DNA. A perfect mirror image can be formed because the structures of uracil and the other bases prohibit them from binding to any other base but their perfectly matched partner. Cytosine can only bind to guanine, and uracil can only bind to adenine.

The RNA chain that corresponds to the gene is now clipped off the DNA, where it migrates to the cell cytoplasm and encounters a ribosome. Ribosomes are tiny, protein-rich structures that form a perfect pocket in which RNA and amino acids can dock. This RNA is called "messenger RNA" because it conveys the DNA's message from the cell nucleus to the cell cytoplasm.

Also entering the ribosomes is yet another type of RNA molecule, called "transfer RNA." These RNAs have at one end of their structure a short sequence of three bases that is complementary to the sequence of a region of the messenger RNA that codes for one, and only one, amino acid. Let's imagine that a messenger RNA molecule leaves the nucleus, with instructions to build a particular protein that happens to have a leucine amino acid (CUG) in its structure. As the messenger RNA reaches the ribosomes, a transfer RNA with a sequence on one side of GAC (the RNA complement to CUG) binds to the messenger RNA. They form a miniature double-stranded RNA molecule along those three bases. At the other end of the transfer RNA, a molecule of leucine is bound. This brings the leucine molecule into the ribosome, where enzymes can link it up with whatever the previous amino acid in the growing protein had been. The transfer RNA leaves, and another takes its place. This one has a complementary code for the next triple-base sequence of messenger RNA, which may correspond to arginine or any other amino acid. In this way, a protein is built up one amino acid at a time, until the entire base sequence of the messenger RNA has been translated.

DNA is sensitive to high energies, like those of radiation. Gamma rays, X rays, cosmic rays, ultraviolet light from the sun, as well as certain drugs and chemicals, can interfere with the sequence of bases in DNA. Such changes are known as mutations and can be as simple as substituting a single base (say, a T) for another (say, G) within the entire 3 billion base sequences. This may change the amino acid code for that tiny strip of DNA and could lead to consequences that range from trivial to lethal. Since DNA replicates itself each time a cell divides, the mutation will persist and will be passed on to offspring. If a mutation results in a disease, such diseases can often be treated but rarely cured. Curing the disease might require correcting the faulty gene so that it no longer produces an abnormal protein. Such a technical feat is on the horizon in science and medicine, but is only in its infancy at this time.

Proteins

Proteins (from the Greek *proteios*, "first," as in first in importance) are produced when a gene is activated by signaling molecules that are generated within cells; thus they are known as gene products. Proteins are involved in countless activities, such as forming the architectural "skeleton" of our bodies, acting as enzymes to initiate chemical reactions, acting as hormones in the brain and bloodstream, and serving as carriers for compounds that don't dissolve well in blood (like fats and oxygen). All living tissue in large measure is made up of various proteins.

All proteins can be divided into two classes, those that do not dissolve in water (fibrous proteins, such as keratin in your fingernails, collagen in your bones), and those that do (globular proteins, such as albumin and antibodies).

Regardless of their solubility in water or their specific functions, all proteins are composed of the same twenty amino acids, albeit in different combinations. An amino acid is a small, carbon- and nitrogen-containing molecule, with a slightly acidic nature. It is the order in which amino acids are strung together (via chemical bonds), known as the *primary* structure of a protein, that determines whether it becomes a hormone, a component of the muscle system, or an antibody. The simplest proteins have only a few amino acids, while larger ones have hundreds.

Certain amino acids repel or attract each other, because, for example, they may contain atoms with positive or negative electric charges. Other amino acids tend to form close "associations" with other amino acids because they may both share similar hydrophobic regions within their structure. These water-fearing regions tend to stick together, much like oil tends to form droplets on the surface of water. When the varous electric and hydrophobic forces occur, the protein folds and twists as some amino acids strain to cluster together, and others try to push one another away. This new, twisted shape is known as the *secondary* structure of the protein, and sometimes resembles the way DNA forms a helix. It is this secondary structure that allows groups of hydrophobic amino acids to clump together. Secondary structure enables proteins to migrate into cell membranes, which—because of the oily nature of cell membranes—prefer water-shunning regions of molecules. By anchoring itself in a cell membrane, a protein can interact with molecules outside the cell (that is, it can act as a receptor, or sensory molecule) and convey information about the outside world to the cell interior.

As a protein assumes a secondary structure, new possibilities arise for interactions among amino acids. For example, two amino acids that may have been separated by 100 intervening amino acids in the linear chain, thus making them too far away to interact, may be much closer together in space once the molecule has been folded into the secondary structure. Thus, when the entire protein molecule has finished folding in on itself, a stable, three-dimensional shape is created that bears no resemblence to the simple, linear array of amino acids that began the process. This level of structure—known as a *tertiary* structure—is extremely important, because it is the 3-D shape of the protein, not a chemical reaction, that allows it to interact only with certain other proteins, like a lock-and-key mechanism. That's why an enzyme that reacts with one molecule will not exert the same actions on all other molecules in the body.

Finally, a fourth, *quaternary,* level of structure exists in which two or more proteins with 3-D structure come together to form a new, larger molecule that is highly stable. An example of such a protein is hemoglobin, the oxygen-shuttling molecule in red blood cells (figure 1).

All these twistings, turnings, attractions, and repulsions depend entirely on the cellular machinery getting the original linear array, or sequence, of the amino acids correct. If a gene contains an error (mutation), or the cell translates the gene's message incorrectly, a protein will not assume its normal shape. The consequences of this can be devastating. A single amino acid error in the hemoglobin molecule, for example, results in the disease known as sickle cell anemia.

Antibodies

The importance of the relationship of a protein's shape to its function is illustrated by the immune system. Other than the brain, the immune system is possibly the single most complex physiological system in the human body. It is composed of cells such as T-cells (*T* because they are produced in the thymus gland), which attack foreign bodies such as bacteria, viruses, and organ transplants. In addition, other kinds of immune cells secrete antibodies into the blood. These antibodies work to sequester foreign proteins out of the circulation—for

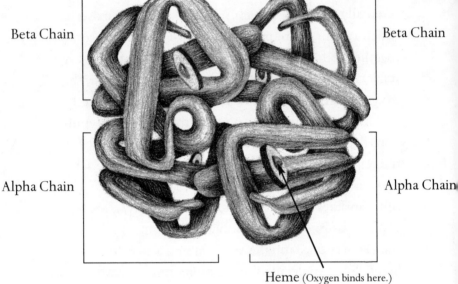

Beta Chain

Beta Chain

Alpha Chain

Alpha Chain

Heme (Oxygen binds here.)

Figure 1. HEMOGLOBIN

Two sets of identical "chains," or subunits, join together to yield one hemoglobin molecule with quaternary structure. Each subunit contains an iron-based heme group that can bind an oxygen molecule. Thus, one molecule of hemoglobin binds four molecules of oxygen.

example, proteins originating from infectious microorganisms.

Antibodies fall into the class of proteins called globulins (because of their water solubility and globular shape) and are given the more specific name immunoglobulins because of their origin in the immune system. The general shape of all antibodies is essentially the same. They consist of four separate chains held together by amino acid "bridges." A special amino acid called cysteine contains within it a sulfur atom connected to a hydrogen atom at one end and a carbon atom at the other end. When two cysteines along the length of a protein are in close proximity due to the tertiary structure of a protein, the hydrogens from each sulfur atom are kicked off, and the two sulfur atoms recombine to form a sulfur-to-sulfur linkage. This bridges one cysteine in the protein chain to another in a different region of the protein, or even one entire protein to another. Both of these situations occur to produce a four-chain antibody.

Antibodies have two heavy chains (so called because they are big) and two light, or small chains, all linked together by cysteine bridges (this gives the antibody its quaternary level of structure). A portion of the amino acids that constitute any antibody are identical in all antibodies. These identical regions play a role in allowing other cells and proteins to find and eliminate the antibody together with its attached foreign protein, regardless of what that foreign protein may be. A second region of all antibodies is unique from one antibody to another. This unique region of the antibody is where binding reactions take place; it is here that the antibody attaches to a specific "antigen" or foreign molecule (for example, a protein released by a bacterium). The variability in this binding region thus allows a wide array of antibodies to be formed, each of which can attack its own particular antigen, and not others.

Sometimes the immune system runs amok. Molecules in an individual's body are incorrectly assessed as "foreign" and attacked by the immune system. Antibodies that mistakenly attack a wide variety of normal tissue proteins have been identified in certain diseases. Such diseases are known as autoimmune disorders because the immune system is attacking one's own self. Type I diabetes, systemic lupus erythematosus, Addison's disease, Graves' disease, and myasthenia gravis are all examples of this insidious breakdown. Sadly, the cause of these breakdowns in immune activity is not currently known.

[2]

Energy: Why Sugar and Fat Aren't All Bad

Just as DNA, RNA, and proteins were necessary for life to have begun, a proper and constant supply of energy to feed cells is required for life to continue. Energy in the body begins as food, which is digested and then absorbed through the gastrointestinal tract. What circulates through the blood, however, is quite different from what was eaten. When we eat a steak, the protein in the steak is chopped into its amino acids, which then travel through the blood and are picked up by cells and converted into proteins again. The fat in the steak is broken down also, eventually stored in adipocytes (the cells that make up our body fat) until needed again. If sweets are eaten after the steak, excess carbohydrates would also be stored, partly as a polymer of sugar called glycogen and partly as fat after conversion of sugar to fatty acids by adipocytes.

The storage of potential fuel sources as fat or glycogen means that people (or any fat-storing animal) can go periods of time without eating. We do not need to eat continuously. If we miss a meal, or even a month of meals, we have enough energy stored as fat, glycogen, and proteins to keep us alive. That's because fat, glycogen, and proteins can be degraded back into smaller, usable molecules as easily as they were originally built up. These small molecules, like glucose (sugar), fatty acids, and certain amino acids serve as fuel to supply the chemical needs of our cells. Each cell in the body has the capacity to "burn" these fuels (or in the case of amino acids, sugars derived from the acids), which provides both heat and a stored form of mechanical energy called ATP (adenosine triphosphate). The heat and the ATP are used as power sources for enzymatic and other chemical reactions.

Sugars and Starch

The most important minute-to-minute fuel in the body is the sugar known as glucose (from the Greek *glykys,* or sweet). Although naturally occurring in certain fruits, most of the glucose we consume comes in the form of more complex sugars, like sucrose (table sugar). One of the key features of sugars is that they can be strung together to form polymers. Sucrose is a very small polymer consisting of one glucose and one fructose linked together, to form what is known as a dissacharide. Another common dissaccharide is milk sugar, or lactose (one glucose plus one related sugar called galactose). Within our intestines are enzymes that clip these dissaccharides into their two smaller component parts, which allows them to more easily be absorbed and shuttled around the bloodstream to hungry cells.

An important aspect of the chemistry of glucose is that it is a small, simple molecule, containing only 6 carbon atoms, 6 oxygen atoms, and 12 hydrogen atoms. This simplicity makes it easy for the energy-producing machinery within our cells to sequentially break down a glucose molecule into smaller and smaller bits. Each time glucose is broken down, energy is produced. That's because it took energy to form the chemical bonds of glucose, and that energy was "stored" in the intact molecule. The energy derived from glucose metabolism takes the form of the compound ATP, which can be saved and used whenever needed for a variety of processes.

The body needs a supply of glucose at all times, because our brains rely almost entirely on glucose for energy production. Other cells in the body can use fats for energy, but the brain must make do on glucose. Therefore, the blood level of glucose cannot be permitted to fall very much, even in a starving person, or brain function would be rapidly compromised. The maintenance of a steady level of serum glucose is one of the greatest achievements of the human body.

When we eat a meal, we usually ingest more energy (food) than is needed for the immediate future. The excess food is deposited partly as fat, but also as another fuel-storage molecule called glycogen (animal starch). Glycogen is a very long polymer of glucose molecules strung together end to end, and stored primarily in the liver (figure 2). When a meal is skipped, an enzyme is activated in the liver that cleaves the glycogen chain into its many individual glucose molecules, which are then transported from the liver into the blood. This process, known as glycogenolysis, is stimulated by hormones like epinephrine (adrenaline). Thus, if a meal or two is skipped, some of the excess fuel stored in the liver as glycogen can supply the body (and, most important, the brain) with a source of glucose for a period of time. Unfortunately, the liver has a finite capacity to store glycogen, and most of it will eventually become depleted within a day or so. At that stage, the only way to continue supplying glucose (without eating) is to convert amino acids (from protein) and glycerol (from triglycerides) into new chemicals that the liver can in turn convert into glucose. This process, known as gluconeogenesis (creation of new glucose), can continue as long as stores of fat and protein last. Hunger strikers have lived for up to two months without food, subsisting only on water and vitamins.

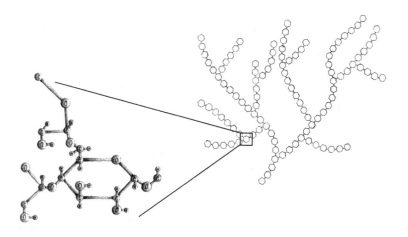

Figure 2. GLYCOGEN

Only a small portion of this starchy polymer is shown. Each hexagon is an individual molecule of glucose, linked end-to-end with its neighbors for efficient packaging in the liver.

Triglycerides

A well-balanced diet has roughly 30 percent of its calories in the form of fat, most of it as triglycerides. Fat serves a number of purposes in the animal kingdom, such as providing an insulating layer of warmth and encasing internal organs for protection. But the most important feature of fat in human beings is its ability to act as an energy source when food is not available.

A triglyceride is composed of a three-carbon molecule of glycerol with a single fatty acid attached to each carbon. The joining of fatty acids with glycerol is reversible; enzymes called lipases can break a single triglyceride back into three fatty acids and one glycerol molecule. When a fatty acid is released from a triglyceride molecule, we call it a "*free* fatty acid."

Should someone be subjected to a prolonged fast, or starvation, carbohydrates stored in the liver would be the first energy source to be tapped. Within a day or so, however, the amount of sugar stored as glycogen in the liver would become depleted. Unless an alternate form of fuel were available, we could not survive long without food. But with a large enough supply of triglycerides, we can hold out for weeks or months without food, provided water is available (although, of course, we would not be particularly healthy or functional).

The more triglycerides packed away in our body, the better able we are to withstand starvation. Unfortunately, the more triglycerides we have, the fatter we are, too. That's because fat cells—adipocytes—are the major place in which triglycerides accumulate, and an adipocyte can enlarge enormously to accommodate massive amounts of triglycerides.

Because triglycerides are large, bulky molecules, they are not easily absorbed through the intestines. Lipases, released by the pancreas into the intestine, are needed to break the triglycerides into fatty acids and monoglycerides, which are smaller and more readily absorbed across the intestinal wall. Once they've gotten inside the intestinal cells, the fatty acids and monoglycerides are repackaged into triglycerides again and released into the bloodstream. Upon reaching an adipocyte, the triglycerides are broken down once again, shuttled across the fat cell membrane, and repackaged all over again into triglycerides. As circuitous as it seems, this process is the only possible way to absorb, transport, and store these large, complex molecules.

During a crisis, such as starvation or stress, hormones activate a special lipase within adipocytes to convert the stored triglycerides back into fatty acids and glycerol, which are then released into the blood. What happens next is rather remarkable. All the cells of the body, save for those in the brain, adapt themselves to use fatty acids—rather than glucose—as their primary fuel source. In fact, the same hormones that activate triglyceride breakdown also inhibit the ability of non-brain cells to use glucose, so that fatty acids must be used in order to survive. Like glucose, fatty acids can be converted into

ATP for energy. The brain does not have the ability to use fatty acids like the rest of the body, however, and thus must use glucose. In addition, those hormones that prevent cells outside the brain from using glucose are ineffective *within* the brain.

So, when fuel is scarce, the brain gets all the glucose it needs, and the rest of the body makes do with fatty acids. And what happens to the glycerol? It circulates to the liver, which can pick it up from the blood and, through yet another biochemical pathway, transform it into a molecule of glucose, which can help continue feeding the brain. This shouldn't be too surprising, since glycerol is a three-carbon molecule, essentially half of a six-carbon molecule of glucose.

Fatty Acids

Fatty acids seem to be the rage nowadays. We hear about unsaturated and saturated fatty acids, *trans* fatty acids, and omega-3 fatty acids. For all their importance, they are really rather simple lipids composed of a string of carbon atoms (usually around 16 to 24 carbons long) with an acid group at one end. The acid group is the same one that is found at the end of amino acids and is called a carboxylic acid. An amino acid is a carboxylic acid with a nitrogen-containing side group at one end called an amino group. A fatty acid is a carboxylic acid attached to a carbon chain, known as a hydrocarbon (because it has only hydrogen and carbon in its chain).

Figure 3. SATURATED FAT

When all the carbon atoms except the terminal one share single bonds with each other and with hydrogen atoms, the molecule is said to be a saturated fatty acid. This particular molecule happens to be palmitic acid, which is obtained from palm oil.

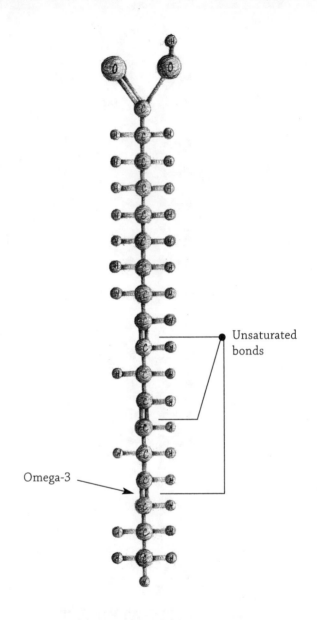

Unsaturated bonds

Omega-3

Figure 4. UNSATURATED FAT

This is a polyunsaturated fatty acid, because several carbon-carbon bonds have "doubled up," due to removal of hydrogen atoms.

If every carbon atom is attached, or bound, to its neighboring carbon atom in the simplest possible way, we say the fatty acid is saturated (figure 3). In other words, each carbon atom (except the ones on the ends of the chain) binds to two other carbons, one on each side, plus two hydrogen atoms. Saturated fatty acids are typically found in animal fat. The most common one is called stearic acid, or sometimes stearate for short.

If some of the hydrogen atoms are removed from neighboring carbons, those carbons will not have all four of their available binding sites occupied, leaving them unsaturated. In that case, the carbons "double up" with each other and form a special kind of bond known as a double bond (figure 4). Unsaturated fatty acids may have one double bond in which two neighboring carbon atoms share electrons with each other, rather than with a hydrogen atom. Such fatty acids are said to be monounsaturated, because only one set of hydrogens is missing. If multiple double bonds occur, the acid is said to be polyunsaturated. Unsaturated fatty acids are typically found in vegetable and fish oils; common ones include oleic acid (oleate), linoleic acid (linoleate), corn oil, and the so-called omega-3 fatty acids. Since the position of the double bond is numbered with respect to the ends of the molecule, an omega-3 fatty acid has a double bond that is three carbons in from the beginning (non-acid part) of a fatty acid. (For an unsaturated fatty acid in a *cis* conformation, see the figure on page 28.) This numbering system helps scientists classify fatty acids.

Omega-3 and omega-6 fatty acids, for instance, have different chemical properties and biological actions.

If the two remaining hydrogens on a double-bonded pair of carbon atoms are both on the same side of the carbons, as is usually the case, the fatty acid is called a *cis* fatty acid; when the hydrogens are on the opposite sides of each other across a double bond, it is called a *trans* fatty acid. The position of the hydrogens in *cis* and *trans* fatty acids is not necessarily permanent. During the processing of margarine, the conformation of some *cis* fatty acids are converted to a *trans* conformation. Although this improves the consistency and stability of margarine, it results in an abnormal lipid whose health consequences are currently debated. Recent evidence suggests that *trans* fatty acids may interfere with the ability of the liver to clear cholesterol from the blood, thus contributing to a higher than normal circulating cholesterol level.

When carbon bonds double up, they become less stable. As a result, unsaturated fats tend to melt more readily (that is, at lower temperatures) and exist as oils (liquids) at room temperature. Saturated fats, on the other hand, are very stable and are usually solid at room temperature. If a saturated fat is heated enough, though, it will melt (the grease in a frying pan after cooking a hamburger). Conversely, if an unsaturated fatty acid is cooled enough, it will form a solid.

Unsaturated fatty acids have been linked with numerous health benefits, including the prevention of cancer, atherosclerosis, high cholesterol, and others, although the mechanism of these protective functions is still unknown. Saturated fats, while not dangerous in normal amounts, are associated with high cholesterol levels and heart disease if eaten in excess. Part of the explanation for the effects of saturated fats on health and disease is that the excess dietary fats are converted by the liver into cholesterol by a complex series of enzymatic reactions. Therefore, a healthy diet should consist not only of reduced amounts of cholesterol, but also less *trans* fatty acids and saturated fats.

Insulin

Of the hormones in the human body, insulin is certainly among the most important. Like many hormones, insulin is a small protein molecule but with a complex and interesting structure. Within the pancreas, insulin is produced as a single long chain of amino acids. The completed molecule is folded back on itself due to chemical bridging between cysteine amino acids. Proper formation and storage of insulin also requires a trace amount of zinc, a positively charged metal atom. This is one reason why a small amount of zinc must be part of a healthy diet. (Zinc deficiencies aren't very common, however, since the mineral is found in such a wide variety of foods, such as meat, eggs, seafood, legumes, milk, grains, nuts, and others.)

Why is insulin so important? Many people may know

that insulin is involved in some way with the regulation of blood sugar. Specifically, insulin is required for sugar (glucose) to traverse the membranes that cover all cells in the body. Cell membranes act as a barrier against water-soluble compounds, preventing them from crossing into and out of the cell interior. Without membranes, our cells would leak their contents and eventually die. Because the membranes work so well, they present an impenetrable barrier for even relatively small molecules like glucose. To get glucose into a cell, a carrier of some sort is needed to latch on to glucose outside the cell, then "flip" across the membrane, dumping the glucose into the cell interior. Insulin is the hormone that causes the production of the glucose carrier molecules. Without insulin, there would be insufficient numbers of glucose carriers, and therefore little or no glucose would be able to cross the cell membrane and enter into cells. Without insulin, therefore, cells starve despite being immersed in a sea of fuel sources. This is what happens in the disease diabetes mellitus.

Insulin has numerous other actions in the body, including the transport of amino acids across membranes and the acceleration of growth and differentiation. It also stimulates fat storage as a means of providing an available source of energy during times of starvation. Most people with diabetes develop the disease in middle age, usually after a period of chronic weight gain and lack of exercise. Those individuals can be distinguished from the smaller number of people who develop diabetes early in life. In the latter case, the disease results from destruction of that part of the pancreas that produces insulin (due to an abnormal immune response). In adults, the disease is usually the result of cells in the body losing their normal sensitivity to insulin, despite the fact that insulin is still being produced by the pancreas.

Treatments for diabetes depend upon which form of the disease is present. If a person has so-called Type 1 diabetes mellitus, insulin must be replaced every day by injection or other means because the person's pancreas cannot make insulin. In the more common form of the disease, Type 2 diabetes mellitus, additional treatments are available. Exercise and weight loss are often sufficient to restore sensitivity of the body's cells to the insulin coming from the pancreas. In more severe cases, a drug can be taken that stimulates the pancreas to oversecrete insulin, the idea being that if cells are less sensitive to insulin, then maybe if much more insulin becomes available to the cells, it will compensate for the decreased sensitivity. If this treatment fails, then insulin must be injected each day to provide sufficient amounts to stimulate the production of glucose carriers.

ATP

All of the fuels discussed in this chapter dissolve in the watery environment of cytoplasm, the fluid-filled space within cells. Although the sugars and fats are the sources of our bodies' energy, they cannot in and of themselves provide useful energy. They can't drive the chemical reactions that are the fundamental processes of life. Instead, the energy needed to drive these reactions comes in the form of a small molecule that is formed in abundance each time a fatty acid or glucose molecule is "burned" within a cell.

Fuel sources in our diet have packed within their structures a form of stored, or potential energy, which derives from the energy it takes to make the chemical bonds between the atoms that make up each fuel molecule. Ultimately, the stored energy of fuel molecules is derived from sunlight: the photons from the sun provide enough energy for plant cells to manufacture their own nutrients. Animals eat the plants, and carnivores eat the plant-eating animals. Thus, the energy gained from sunlight is transmitted up the food chain. The ability to recover that energy and store it in a chemical form that can be tapped whenever needed is vital to the survival of all animals.

The stored energy takes the form of a molecule called adenosine triphosphate, or ATP for short. It is a rather unusual molecule, composed of a sugar (ribose), a nitrogen-rich base called adenine, and a string of three phosphate molecules.

The ribose-adenine backbone of ATP is known as adenosine and is widely found in plant and animal tissues, and is therefore part of our usual diet. When carbohydrates or fats are burned within the specialized cellular structures called mitochondria, the bonds between the carbon, oxygen, and hydrogen atoms are broken. This releases some of the pent-up energy that was stored in those bonds when the molecules were first formed. Some of the released energy dissipates as heat; it is partly this heat that allows us to keep warm even on a cold day. Another portion of the energy, however, is recaptured in another chemical bond, that of ATP. Phosphate groups are added to adenosine. In essence, the energy from sunlight is stored in fuels, and the energy of the fuels is then redirected to chemical bonds in ATP. This is very convenient, because the phosphate bonds in ATP can be easily broken, releasing their stored energy in the process. Thus, all cells keep a ready supply of abundant energy in their cytoplasm in the form of ATP.

What is the energy in ATP used for? One example that occurs constantly is the enzyme-dependent conversion of one molecule into another. Another common example is the powering of actin and myosin fibers within muscle cells (as we'll see in more detail in chapter 6). Yet another is the power required to operate the cellular "pumps" that maintain a proper electrolyte balance within and around our cells. It's a simple sequence: sunlight, plants, herbivores, carnivores, breakdown of chemical bonds in fuels, recapture of energy in ATP.

A final note in the ATP story: if the last two phosphate groups are clipped off ATP, the result is a unique molecule known as adenosine *mono*phosphate (because it has only one phosphate left). The single phosphate group, without another phosphate to bond with, "cyclizes," that is, it bonds with two regions of the ribose part of AMP. This new molecule, known as cyclic AMP, is of extreme importance. It acts as the link between signals reaching the outside of cells and the cell interior. For example, when a hormone activates a receptor molecule on the cell's membrane, cyclic AMP is formed inside the cell by an enzyme that chops up ATP. The cyclic AMP then diffuses through the cell and initiates a host of activities vital for the survival of the cell and the organism. Its discovery justly earned a Nobel Prize for Earl Sutherland in the 1950s.

[3]

Digestion: The Good, the Bad, and the Ugly

Now that we have seen how the fuels that provide energy to the body's cells are stored and used, we'll take a look at how fuels and other molecules are first digested and absorbed. We will also look at cholesterol, an important molecule with a complex role in normal health—but one that can become extremely unhealthy in excess.

Most of the work of digestion requires the presence of enzymes. Enzymes are common proteins, but they are unique in that they can latch on to a specific molecule, such as another protein, and cut that molecule into smaller fragments. Elsewhere in the body enzymes may do the opposite, namely build large molecules from smaller ones. But in the intestines, enzymes generally break apart large molecules so that they can be more easily transported (absorbed) across the cells of the intestines and into the bloodstream. Thus, a meal that contains protein, triglycerides, and complex sugars (dissacharides) is degraded into amino acids, fatty acids, and simple sugars, respectively. These smaller molecules are then absorbed and transported through the blood to places where they are needed. It is precisely for this reason that individuals who take insulin to control their diabetes, as opposed to controlling the disease by diet, exercise, or drugs, must inject the insulin. If the insulin were taken orally, the enzymes in the intestine would degrade the insulin into its component amino acids, which would then be absorbed into the blood and shuttled around the body to be converted into other proteins. In the near future, however, inhaled and nasally administered insulin sprays should become available that will alleviate the need for injections in some people.

There are, however, some molecules that are already relatively small and do not need to be enzymatically degraded prior to their absorption. One such molecule is cholesterol, which, because of its importance in human disease, deserves special mention.

Cholesterol

Cholesterol is commonly regarded as one of the great dietary evils of modern times. Too much cholesterol can indeed be harmful, but surprisingly, we could not sur-

vive with too little cholesterol, either, because it is a component of cell membranes and a precursor to other vital molecules.

Cholesterol is a lipid; that is, it easily dissolves in oil but does so only poorly in water. It has a structure made up of several "rings" of carbon atoms (figure 5). Because cholesterol is a lipid, it moves easily across the oil-rich cell membranes of the intestine. But because cholesterol does not dissolve well in water and blood is mostly water, cholesterol doesn't dissolve well in blood. And therein lies a problem. In order to be transported in the watery blood, cholesterol must be properly packaged or it will fail to be fully dissolved and will precipitate as a solid in the blood. When cholesterol precipitates in blood, it tends to latch on to the walls of blood vessels and form the nucleus of what will become an athero-sclerotic plaque. In fact, advanced plaques are actually composed largely of blood vessel muscle cells, but these are believed to be stimulated to multiply by the presence of cholesterol deposits and other factors. As a plaque enlarges, it can occlude a blood vessel so severely that the amount of blood getting through the narrowed space is reduced, and the cells of the body supplied by those vessels become deprived of oxygen and nutrients. If this happens in the arteries of the heart (which, for unknown reasons, are particularly susceptible to plaque formation), heart cells can die, which leads to heart failure.

To cart cholesterol around in the blood, most of it is

incorporated into large, bulky structures called lipoproteins. As their name implies, lipoproteins are composed of lipids and protein. The protein, which surrounds the cholesterol and readily dissolves in water, acts as a means to prevent cholesterol from precipitating, and also as a sort of taxi to shuttle cholesterol to the places it's needed.

(Not all hydrogen atoms are shown.)

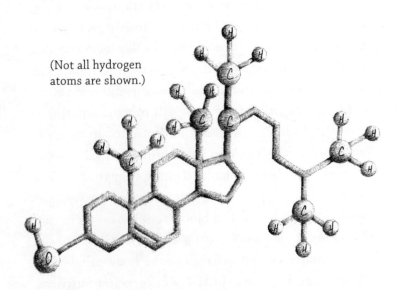

Figure 5. CHOLESTEROL

The ring structures, shaped like hexagons or pentagons, are made up of six and five carbon atoms bound together.

In one type of lipoprotein, called high density lipoprotein (HDL), the amount of cholesterol and other lipids packaged with the protein is relatively low. Women often have higher levels of HDL than do men, partly because estrogen stimulates HDL formation, whereas testosterone inhibits its formation.

In another lipoprotein, called low-density lipoprotein (LDL), the ratio of cholesterol to protein is high (giving the particle a low density since fat is less dense than water; in other words, fat floats). HDLs bring their cholesterol to the liver, where it can be removed from the circulation; thus, this type of cholesterol is called "good cholesterol." LDLs are also removed from the blood to some extent by the liver, but in high amounts they tend to accumulate in the circulation. It is the cholesterol in LDLs that forms plaque; that is why LDL is known as "bad cholesterol." In fact, the cholesterol in both HDL and LDL is chemically identical.

You may wonder why cholesterol is needed at all if its presence in the blood can cause such problems. There are at least three major reasons. First, cholesterol is an important part of all cell membranes; without it, cell membranes take on a liquid-like characteristic that impairs cell function. Second, cholesterol is used to make bile salts in the liver, which are needed for digestion of fats. And last, but not least, cholesterol forms the backbone from which the steroid hormones in the body, such as cortisol, aldosterone, testosterone, progesterone, and estrogen, are made.

Bile

In the liver, some cholesterol is sequestered from the bloodstream and converted into bile. The bile is secreted in the gallbladder, where it is stored until the next meal comes along. When that future meal is eaten, the bile is released from the gallbladder into the intestine, where it helps to digest fatty foods.

Bile is not a single molecule but actually a mixture of molecules, including electrolytes, pigments, electrically charged lipids called phospholipids, and molecules derived from cholesterol called bile acids. The pigments are yellow-colored breakdown products of bilirubin, which is itself a breakdown product of hemoglobin from old red blood cells. The pigments are not involved in the digestion process and are therefore eliminated in the feces. Thus, as red blood cells age and die, the hemoglobin ends up as bilirubin in the bile.

The major bile acids are two substances called cholic acid and chenodeoxycholic acid, both of which are made from cholesterol. The liver extracts cholesterol from the blood, converts it into these two bile acids, and sends them on their way to the gallbladder through a duct that connects the two organs. When a meal containing fat is ingested, special cells in the intestine sense the presence of fat and release a hormone called cholecystokinin (CCK) into the bloodstream. CCK eventually reaches the gallbladder, where it stimulates contraction of the gallbladder wall. This squeezes the bile out through another

duct into the intestine, where it mixes with the ingested food. The bile does not flow back to the liver (where it originated) but ends up entirely in the intestine.

Bile acids are a bit strange in that they are partly acid and partly lipid. Although they resemble cholesterol in structure, they have a carboxylic acid group at one end. This gives them the unusual property of having one part (the acid end) soluble in water and the other part (the end that most resembles cholesterol) soluble in oil.

So when bile mixes with fat and other foods in the intestine, the oily parts of bile immerse themselves into fat droplets, while the watery parts of bile stick out of the droplet and form a sort of watery coat. Incidentally, bile is not the only substance that can be soluble in both water and oil at different sites within its structure. Soap works the same way. A fat-soluble region of soap soaks up oily droplets on our skin, and a water-soluble region of the soap molecules allows it to be washed away by water.

As the intestines churn up the partially digested food, big fat droplets break into smaller ones, just as a layer of salad oil would break into smaller oil droplets if shaken in a glass of water. This allows more surface area for fat-digesting enzymes (lipases) to act upon and facilitates the digestion of the fats that we eat. However, it is common knowledge that when you allow a container of oil and water to settle again, the tiny beads of oil will quickly coalesce again into larger ones. This would also happen to the tiny fat droplets in the intestine if it were not for the actions of bile. Because all of the water-soluble, acid ends of bile contain identical electric charges, the surfaces of the small lipid droplets repel one another. Thus, digestive enzymes are not faced with the task of slowly digesting large spherical oil droplets, but instead can act more efficiently on countless miniature droplets.

If a person's gallbladder is removed, bile can still be produced by the liver, but it will not be released into the intestine at the right times, because CCK would then have no storage reservoir upon which to act. Such individuals often eat a diet low in fat, or consume many small meals daily rather than three large ones, so that their intestines can handle the fat. Otherwise, a syndrome known as malabsorption may occur, and the undigested fat remains in the intestine causing gas, diarrhea, and loss of fat-soluble vitamins ingested in the diet.

About 95 percent of the bile secreted by the gallbladder is efficiently reabsorbed by the intestine after it has done its job. It is then recycled to the liver and gallbladder. Fiber, which is found in plant products, especially legumes, is indigestible and tends to inhibit the reabsorption of bile from the intestine. Consequently, the liver is left with a shortage of bile and must extract more cholesterol from the blood to regenerate the bile being lost in the feces. That's one reason why it's believed a fiber-rich diet has the effect of lowering blood cholesterol levels in a harmless, natural way.

Mucus

Mucus may not be one of the more glamorous molecules in the body, but it certainly is important. Mucus is made up of a collection of molecules called mucins, together with some bicarbonate ions, all dissolved in water. A mucin is a protein made up of four subunits, joined together by sulfide linkages between cysteine amino acids. In addition, these proteins have sugar molecules attached to them and are thus called glycoproteins.

We all know that mucus is in our noses, especially when we have colds. The mucus in the nose serves two purposes. First, it is sticky enough to act as a filter to absorb some of the potentially hazardous dust and debris in the air we inhale. Instead of ending up in the sensitive air sacs of our lungs, airborne debris gets swallowed or blown out. Second, mucus coats the lining of the sensory region of the nose, where the smell receptors are located. Without mucus, there would be no medium for odor molecules to dissolve in and interact with odor detectors on sensory cells. No mucus, no sense of smell.

But mucus is found in many other parts of the body as well and plays extremely important roles in normal health and digestion. Because it is fairly viscous and gel-like, it protects the lining of our stomachs and intestines from the harsh acids secreted by gastrointestinal cells during digestion. If the mucous layer wears away, the acids begin destroying the stomach and intestinal linings, resulting in an ulcer.

Mucins are also part of what makes up saliva; to appreciate how important mucus is in this regard, imagine trying to swallow a dry piece of cake without any saliva for lubrication (think of "cotton mouth"). Without saliva, our teeth would eventually fall out due to decay. This appears to result from antibacterial actions of mucins in saliva. Recent evidence suggests that mucins contain receptor-like characteristics within their structure. These receptors bind and sequester bacteria that grow on teeth and at the same time sequester white blood cells that have bacteria-fighting capabilities. The research suggests that mucins may be important in dental health by acting as a mediator of the immune system.

Mucus also helps with the movement of undigested food through the colon, and it coats the lower end of the esophagus where it joins the stomach. This helps minimize damage to the esophagus from occasional reflux of stomach contents (heartburn).

Overproduction of mucus can be as serious, or more so, than underproduction. Many pulmonary diseases, for example, are associated with excess mucus production in the airways. In such diseases, like cystic fibrosis, asthma, and bronchitis, the mucus accumulates in the tiniest airways, clogging them up and making it harder for air to get through.

[4]

Salt and Water:
Keeping a Balance

All life, from plants to mammals, is water-based. It is in the water of our blood and our cells that chemicals dissolve so that they can interact to form new chemicals. It is in water that hormones are shuttled around the body to exert their widespread effects. It is water that evaporates off our skin, drawing body heat with it and cooling us down on hot days. And it is water that carts off bodily wastes to the kidneys, so that they don't accumulate to toxic levels.

If you weigh 150 pounds, roughly 90 pounds of that weight is water. Those 90 pounds are distributed in three separate compartments: the blood, inside cells, and in the spaces between the blood vessels and the cell interiors.

Most of the water, about 67 percent or so, is in the cells, while only about 8 percent is in the blood and the remaining 25 percent is in between.

In addition to water, our blood is made up of red and white cells and clot-producing platelets. It is the red and white blood cells that give blood its viscosity ("blood is thicker than water"). Dissolved in the watery part of blood are all the nutrients, gases, and other molecules we need to survive, such as vitamins, minerals, hormones, oxygen, amino acids, sugars, fats, and salts. Collectively, the water and all its dissolved molecules are known as plasma.

Like cholesterol, salt has gotten a somewhat bad rap. It's true that excess salt has been tentatively linked with high blood pressure, but the jury is still out on the strength of this correlation. On the other hand, a normal balance of salt in our blood is critical for survival. In fact, too little salt in the blood can be just as dangerous as too much. Indeed, salt is somewhat of a tricky word: to the average person it means table salt, but to a chemist it is any acid where the hydrogen has been replaced by a metal atom. In the case of table salt, the hydrogen of hydrochloric acid (HCl) has been replaced with sodium to form sodium chloride (NaCl). But potassium chloride is also a salt (in this case, the hydrogen is replaced by an atom of potassium), and it is also critical for the body's workings. We use salts for therapeutic purposes as well. If the hydrogen in sulfuric acid is replaced by magnesium, the result is magnesium sulfate (epsom salts), a laxative. And if two hydrogens from carbonic acid are replaced by a single calcium atom, we get calcium carbonate, an antacid.

The importance of salt for survival depends upon its properties once it is dissolved in water. At that time, salts break up into the electrically charged particles of the metal and the original acid—thus the term *electro*lytes. As we'll see, it is the electrical nature of salts that contributes to their ability to regulate brain and heart cell activity.

Salt and Water

Water is a very stable molecule made up of one oxygen and two hydrogen atoms. Because oxygen has a strong tendency to "steal" electrons from other atoms, the electrons belonging to the two hydrogen atoms tend to migrate closer to the oxygen atom than to their original hydrogens. This interesting arrangement means that the oxygen picks up a slight negative charge (since electrons are negatively charged). The hydrogens, on the other hand, become slightly positively charged due to the drifting of their electrons. Thus, water molecules are slightly negative at one end and slightly positive at the other. A molecule with two different charges on it at either end is called a dipole, and the dipole nature of water has surprising significance in allowing water to act as a solvent for salts.

Many of the most important molecules in the blood

exist as salts. A salt is a molecule that will dissolve in water, and in doing so will create two molecules where only one existed before. A common example is table salt, which is composed of a single sodium atom joined to a single chlorine atom. When bound together, an electron from the outer shell of sodium "jumps" to the outer shell of chlorine. Like oxygen, chlorine has a tendency to attract electrons, but in this case the electron doesn't merely drift nearer to chlorine, it actually moves completely into the chlorine shell. When dissolved in water, however, the two atoms—sodium and chlorine—break apart from each other. What's left is a positively charged sodium atom, and a negatively charged chlorine atom. Charged atoms in solution are known as ions. Here's where the importance of the dipole nature of water comes into play. Dissolved, positively charged sodiums are attracted to the negativity of the oxygen atoms of water molecules and repelled by the slightly positive hydrogen atoms. The reverse is true for the negatively charged chloride atoms. Thus, the sodiums and chlorides not only break apart but remain apart because they are surrounded by the electrical charges of water molecules (figure 6).

Ions are necessary to drive the electrical events in the heart and brain. Anytime an ion, or any other charged molecule for that matter, moves through space it creates an electric current. Without water, no ions. Without ions, no electrical signals in the brain and heart. The electric current that drives our heartbeat and allows communication between brain cells, then, depends on a precisely regulated salt concentration. That means that a decrease in the water content in the body (dehydration) or an increase or decrease in the amount of salt consumed, will change the salt *concentration* of the blood. You can have a normal number of salt molecules, but if the fluid content of the body drops, the concentration of the salt increases. This will skew the electric currents flowing back and forth across cellular membranes in the brain and heart. The result can be catastrophic. For example, in people with certain hormonal imbalances that cause diuresis (excessive water loss in the urine) and therefore a loss of body fluid, seizures may result due to the ensuing electrolyte imbalance.

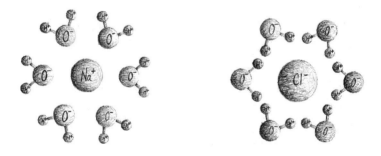

Figure 6. SODIUM CHLORIDE IN WATER
A molecule of sodium chloride (NaCl; table salt) dissolves in water to yield two charged ions. The dipole nature of water molecules prevents the two ions from re-forming into NaCl.

The dipole nature of water also makes the molecule exceedingly stable. Each water molecule attracts neighboring water molecules above, below, and to the sides of it. Positive regions attract negative regions, and vice versa. Thus, it takes a lot of energy to alter the physical state of water from its stable liquid form. In fact, water is about the only liquid that remains a liquid at the full range of temperatures found in living organisms. We are in no danger of having our blood boil, an event that would only occur at greater than a hundred degrees above the highest temperature our blood could ever reach.

A curious event occurs at the extreme surface of a water layer. There is no water molecule above the topmost water, so the top layer of molecules are attracted by fellow molecules to the sides and below, only. Without the molecules above it, there is a net attraction inward and down, which tends to make water form droplets. This phenomenon is known as surface tension and is extremely important in the lungs of air-breathing animals, particularly at the moment of birth.

Incidentally, water is much more than a medium for dissolving salts and other substances. It also participates in chemical reactions. For example, to break down the protein molecules in a mouthful of steak, enzymes must add a water molecule at each place where one amino acid joins another. This breaks the bonds between the amino acids and allows the protein to be reduced to smaller units that are more readily absorbed into the blood.

Angiotensin

I once witnessed an experiment where a few billionths of a gram of angiotensin was injected into the brain of a dog that had just finished drinking its fill. Within moments, the dog showed all the signs of an animal stranded in a desert. That is, it developed a ferocious thirst and started drinking all over again. That's because angiotensin is a peptide hormone known as a dipsogen, which is anything that makes an animal or person thirsty.

Dramatic example aside, the dipsogen nature of angiotensin may really be one of its least important roles. It is also one of the most potent molecules known in raising blood pressure. It does this by stimulating the contraction of the muscle cells that surround small arteries in the body, which has the same effect as placing a clamp around a garden hose, raising the water pressure behind the clamp.

As powerful as angiotensin is, it starts out as something entirely inactive—angiotensinogen. Angiotensinogen is a large protein that is continually pumped out by the liver. It circulates in the blood as a sort of storage form of angiotensin, ready to be converted into angiotensin at a moment's notice.

Why should we need a readily available source of angiotensin precursor in the blood? The main reason is that it allows us to immediately adapt to a sudden loss of blood pressure. Let's take an example of a person who is losing blood due to some severe injury. The loss of blood causes a drop in blood pressure, which must be corrected or the person could suffer brain damage or even die. Quickly, blood pressure detectors in the major arteries send signals to the brain, which relays the signals to the kidneys. The kidneys contain within them special hormone-secreting cells. These cells respond to the neuronal signals by releasing a hormone called renin into the blood. When renin encounters an angiotensinogen molecule, it clips almost all of the tail end of molecule off, leaving only a string of ten amino acids called angiotensin 1. This molecule is an intermediate with little or no biological activity. It must be further clipped down to eight amino acids, resulting in what is called angiotensin 2. This is the active form of angiotensin, which then acts on blood vessel muscles as described above, to help restore blood pressure.

The enzyme responsible for the conversion of angiotensin 1 to 2 is known, appropriately enough, as angiotensin-converting enzyme (or ACE for short), and it is located in the lining of blood vessels. Based on the knowledge of the effects of angiotensin 2 on blood pressure, scientists have developed drugs that block the action of ACE in people with high blood pressure. By blocking ACE (these drugs are known as ACE-inhibitors), angiotensin 1 cannot be converted as readily to angiotensin 2, and the blood level of angiotensin 2 decreases. This tends to cause a mild reduction in blood pressure and is extremely effective in combatting mild hypertension.

Having a circulating precursor of angiotensin 2 readily available at all times means that only two quick reactions must occur to form the functional hormone, and possibly save a life. It is an efficient means of keeping a potential source of angiotensin 2 handy, without running the risk of having excess angiotensin 2 around when blood pressure is normal.

Angiotensin 2 has one other important action. It stimulates cells in the adrenal glands to make a steroid hormone called aldosterone. "Aldo," as it's often called, acts to increase the amount of salt and water retained by the kidneys. The extra salt and water helps to restore body fluid levels when they are low—because of dehydration, for example, or the hemorrhage example just described.

[5]

Gases in a Sea of Water

The human body is full of a variety of gases. Some are produced by bacteria in our intestines, while others are produced by cells in the brain and are used as signaling molecules, and some are not made in the body at all but are inhaled.

Most gases are simple molecules. Nonetheless, the fact that they exist in a different physical state from that of the rest of the body, which is entirely liquid and solid, poses some interesting challenges. For example, the carbon dioxide (CO_2) which is constantly being produced by our metabolizing cells is the same as the CO_2 used to produce carbonation in champagne and soft drinks. But it wouldn't do for us to be fizzing like a can of shaken soda, so the CO_2 in our bodies needs to be converted into something else. It happens that most of the CO_2 we produce is converted, by enzymes, into the harmless chemical bicarbonate. Likewise, the oxygen we breathe can only be of use to our cells if it is physically dissolved in solution. If instead it is floating through the blood as an air bubble, it cannot enter the interior of cells, where it is needed. Like a bit of solid salt, oxygen must be able to dissolve in plasma. Unfortunately, only the tiniest amount of oxygen can dissolve in plasma, so another mechanism is needed to shuttle the gas to and fro in the blood. That mechanism is provided by the protein hemoglobin, discussed in this chapter.

Obtaining the oxygen we need is no easy chore, either, even though most of us are rarely conscious of the workings of our lungs. Getting sufficient oxygen across the lung tissue and into the blood requires a complex set of neuronal and muscular responses. In addition, the lungs must be kept from deflating fully. Like a balloon, they have a natural tendency to deflate, and would do so were it not for the presence of a substance known as surfactant. This is an extremely important molecule, worth its own discussion in this chapter, because it is responsible for the majority of problems associated with premature births.

The Nobel Prize in Physiology and Medicine was awarded in 1998 to three scientists who discovered that the body's own cells can produce another type of gas, known as nitric oxide. This molecule has been demonstrated to function as a communicator within cells and to exert actions ranging from visual processing to the dilation of blood vessels. This remarkable discovery, that a gas could be generated in the body and used to carry out such diverse functions, merits inclusion in our select survey of some of the most important molecules in the human body.

Oxygen

We are all well aware of our vital need for oxygen, but what exactly does oxygen do that's so crucial? Besides being a component of water, oxygen is used to soak up the electrons that are shuttled around in the cells of the body during respiration. Although we often think of respiration as inhaling and exhaling, it really means the burning of fuel in the presence of oxygen. Sometimes we use the terms "internal respiration" and "external respiration" to differentiate between the combustion events in our cells and the actual mechanical process of breathing.

Oxygen is able to soak up stray electrons because, unlike many other atoms, it has room for two additional electrons around its nucleus. When oxygen "captures" the electrons from other atoms, we say that it has oxidized the other compound. Oxygen is not the only atom capable of oxidizing other compounds, but it's one of the most prevalent. Oxygen atoms even share electrons with other oxygen atoms, forming the molecule O_2.

How does this electron-capturing function of oxygen

relate to respiration? When a fuel such as sugar is metabolized, it must be burned by special energy-producing units found in all cells, called mitochondria. To burn efficiently, oxygen is required. This is why a dying fire burns more brightly if we blow on it; the oxygen in our breath (only part of the O_2 that is inhaled gets used up, the rest is exhaled) facilitates the combustion. This of course doesn't mean we have microscopic fires burning in our cells. Enzymes are present to ensure that the combustion of fuel is tightly controlled, generating heat, not fire!

As the combustion proceeds, electrons are bounced from one enzyme to another, finally reaching the last enzyme in a long chain. This transfer of electrons is used to generate energy by contributing to the formation of ATP, which is the major chemical storage form of energy in all cells. But when the electron reaches the final enzyme, it must be removed or the chain reaction will not be able to start over again. Oxygen, because of its strong tendency to capture and hold on to electrons, thus permits the energy-producing reactions to continue. The newly charged oxygen then attracts and combines with positively charged hydrogens that have lost their electrons and is converted to water. The by-products of respiration, therefore, are heat, ATP, and water. In addition, carbon dioxide is liberated during the reaction.

Oxygen also plays an important role in numerous other chemical reactions, such as the synthesis of steroid hormones and the deactivation of toxic compounds by the liver. It can also bind to iron atoms, which are at the core of hemoglobin molecules in the blood.

Oddly, too much oxygen can be dangerous, resulting in the production of highly reactive compounds like superoxides and free radicals. Free radicals have been postulated to accelerate the aging process. In addition, oxidation of certain lipids, like low-density lipoproteins, has been implicated in the development of atherosclerosis. To combat the deleterious actions of oxygen, therefore, the body requires antioxidants in the form of vitamins like vitamins C (ascorbic acid) and E. These compounds have the ability to reverse oxidation by "donating" some of their own electrons to other molecules.

Under certain conditions, like those existing in the upper atmosphere, three oxygen atoms can combine to create the molecule O_3, or ozone. Ozone is capable of absorbing ultraviolet light, and thus acts as a natural sunscreen for all of earth's surface dwellers.

Hemoglobin

Only a very small amount of oxygen can dissolve in the watery plasma portion of blood, providing too little to support even the minimal energy demands of sleep. Thus, a carrier of some sort is needed to create a reservoir of the oxygen gas. This reservoir can be tapped to supply the energy needs of the entire body, from inactive parts like skin cells to hardworking parts like muscles.

Hemoglobin, a large, bulky protein packaged in red blood cells, is the body's reservoir of oxygen.

To make one molecule of hemoglobin, four separate proteins, or subunits, must be joined together. Each subunit has at its core a chemical group called a heme, which contains a single atom of iron (figure 7). It is the iron that binds oxygen. A single hemoglobin molecule can bind to four oxygens—a most efficient way to transport the gas.

Although iron binds tightly to the oxygen, the reaction is not irreversible. If it were, the oxygen picked up in the lungs would never be released to the rest of the body. Instead, oxygen binds to iron wherever oxygen concentrations are high (the lungs) and comes free of the iron where oxygen levels are low (in active cells that are using up oxygen). To visualize how this works, imagine a simple activity such as climbing a flight of stairs. As you climb, blood passes through the capillaries of the calf muscles of your legs. But as the muscles work harder, they need more oxygen to make energy, and the oxygen within the calf muscle cells is quickly depleted. The oxygen dissolved in the plasma circulating within capillaries (high O_2 concentration) is now driven to diffuse across the capillary and enter the muscle cell (where O_2 concentration is low) to replace the oxygen consumed by the cell. As this happens, the amount of oxygen dissolved in the plasma is in turn decreased, making it easier for oxygen that is bound to hemoglobin to be released. The oxygen molecules that are released

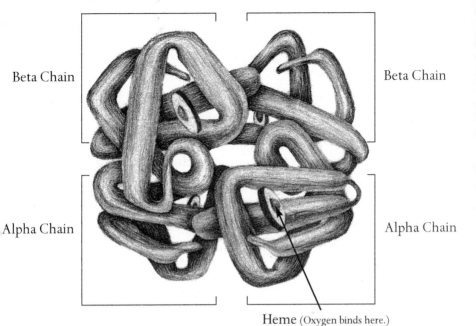

Beta Chain

Beta Chain

Alpha Chain

Alpha Chain

Heme (Oxygen binds here.)

HEMOGLOBIN

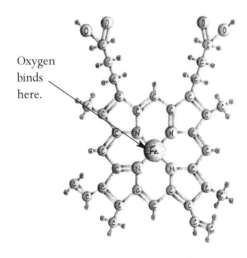

Oxygen
binds
here.

Figure 7. HEME

The three-dimensional structure of hemoglobin is
shown with the expanded structure of a heme group.
One oxygen molecule binds to each iron atom.

from hemoglobin enter the plasma and become available for more muscle activity.

Remember, plasma dissolves only a small amount of oxygen; as the dissolved oxygen leaves the plasma and is used by cells, there is more room for oxygen to be released from hemoglobin and enter the plasma. The newly replenished plasma now has more oxygen to send to the nearby cells, and as long as the cells keep using up their oxygen, this one-way flow of oxygen from hemoglobin to plasma to muscle cell will continue.

Like other large proteins, hemoglobin exists in a precise three-dimensional orientation. Of the 550 or so amino acids that make up this protein, the majority are in place solely to hold the heme group (the iron) and allow it to be accessible to oxygen. If any of the amino acids in the protein chain are altered due to a mutation, for example, then the entire shape of the molecule is changed.

One well-known example of a single amino acid mutation, where a different amino acid is substituted for the normal one, is sickle cell anemia. In that disease, the hemoglobin molecule changes its 3-D shape and becomes less soluble in the watery interior of a blood cell. As a consequence, some of the abnormal hemoglobin forms long, fibrous strands that can no longer be dissolved in the plasma. The strands are said to precipitate out of solution (form a solid). Since red blood cells are little more than bags of hemoglobin, the stiff, fibrous hemoglobin molecules tend to deform the blood cell and make it harder for hemoglobin to properly bind oxygen. The deformed (sickle-shaped) blood cell also has a hard time squeezing through tight capillaries, and the result is often crippling pain and anemia.

Hemoglobin has a second function besides loading and unloading oxygen in the blood. It can also act as an important acid-base buffer in the blood. It does this by soaking up stray hydrogen ions (which are acidic) to help maintain normal pH balance. Hemoglobin is also capable of binding to carbon dioxide (CO_2) and carbon monoxide (CO). Unfortunately, it binds to carbon monoxide even more avidly than it does to oxygen. Once bound, the carbon monoxide prevents oxygen from binding, and the result is a life-threatening reduction in the total amount of oxygen carried in the blood. Carbon monoxide is all around us, in cigarette smoke, in car exhaust, and elsewhere, but does not normally reach levels that are harmful.

If you could somehow increase the amount of hemoglobin in your blood, you would have a greater ability to carry oxygen throughout your body. For a sedentary person, the effect of having some extra oxygen in the blood would not be noticeable. But for an elite athlete competing in highly aerobic activities, extra oxygen means extra performance. Not surprisingly, therefore, some athletes have been known to pack their blood with extra red blood cells (and, therefore, extra hemoglobin). This practice is often called "doping," and it is done with injections of a hormone called erythropoeitin. EPO, as it's known, is a natural hormone that increases our red blood cell count when we climb to high altitude (where oxygen is scarce) or suffer from anemia. When given to a healthy individual, though, the blood cell count can climb so high that it puts excessive strain on the heart. The blood becomes sort of "sludgy." In recent years, several elite bicyclists around the world who were using EPO died of coronary arrest at very young ages. It's a high price to pay just to win a race.

Surfactant

There is only one major internal structure in the body in which air and water come together to form an interface, and it's in the lungs. Within the lungs are tiny air sacs called alveoli, across which oxygen moves into nearby blood capillaries. Like all cells inside the body, the cells of the alveoli are coated with a thin film of water, which prevents the alveoli from drying out. At the surface of this film, however, water molecules tend to attract each other sideways and downward, due to the electrostatic attractions of neighboring oxygens and hydrogens. As a result, the film of water, in contact with air, tugs on the air sacs and forces them to deflate. In a sense, they begin to implode.

Fortunately, a mechanism in all air-breathing animals counteracts the surface tension, as it's called, that exists on the inner surface of the hundreds of millions of alveoli. That mechanism comes in the form of a compound called surfactant. In reality, surfactant is a mixture of several different molecules, including proteins, lipids, and ions, that act collectively to reduce surface tension.

Molecules of surfactant are produced by special cells in the alveoli that are not involved in the exchange of oxygen. These cells secrete the surfactant into the alveolar air space, where the molecules find their way into the water film that coats the inside of alveoli. Surfactant has detergent-like properties; that is, one end is oily and the other is soluble in water. Thus, the molecules orient themselves vertically in the water layer. The oily ends stick up into the air sac itself and provide a shield between the air and the alveolus cell. Each molecule of surfactant squeezes between surface water molecules, increasing the distance between them. This has the benefit of decreasing the attractive force from an oxygen atom in one water molecule to the hydrogen atom in its neighbor, and the surface tension is reduced.

The system doesn't work perfectly, however, and eventually the surface tension—though reduced—overcomes the surfactant and the alveoli start collapsing. Fortunately, all it takes to reinflate them is a nice, deep breath. This occurs automatically if we are active, but even at rest we occasionally sigh, take a deeper breath than normal, or yawn. Such deep breathing forces the lungs open again, since our chest muscles are stronger than the surface tension force. Once reinflated to their normal size, the gradual shrinkage of each alveoli begins again, slowed—but not stopped—by surfactant.

A lack of surfactant has always been the leading cause of disease and death in premature infants. That's because during fetal life, the lungs are filled with amniotic fluid; there is no air to create a water-air interface, and thus no surface tension exists. Indeed, the fetus doesn't need its lungs at all, since it gets its oxygen from its mother through the umbilical cord. Just before birth, the fetus begins making surfactant in anticipation of the arrival of surface tension in its airways. If the fetus is born too early (say, before 7.5 months of pregnancy), surfactant may not be available and severe respiratory problems may arise (known as "respiratory distress syndrome of the newborn"). The alveoli collapse due to the unchecked surface tension, and breathing becomes exceedingly difficult, especially for a weak infant (think how much harder it is to inflate a completely deflated balloon than one already partly inflated). Nowadays, synthetic surfactants are available that provide some relief. In addition, if it is known ahead of time that the delivery of the baby will be premature, the mother is sometimes injected with a steroid that mimics the actions of the natural adrenal steroid hormones. This is done because adrenal steroids stimulate surfactant production and will enter the fetal circulation if injected into the mother. Incidentally, adults can get a form of respiratory distress syndrome, too: cigarette smoke tends to reduce surfactant content in the lungs.

Nitric Oxide

In recent years, it has been learned that the human body manufactures gases other than carbon dioxide and that some of these gases, unlike CO_2, may not be mere waste products of metabolism. Nitric oxide (NO) is one such gas. Chemically, it is one of the simpler molecules in the body, although the process by which it is formed is enormously complex. In essence, a solid (the amino acid, arginine) is converted into a gas. That may not sound like much, but think of how much heat is needed to convert liquid water in a teakettle to water vapor. Cells in the body, though, can convert arginine to gaseous NO using enzymes that facilitate the reaction.

Since the reaction requires an enzyme, it only occurs in those parts of the body where the enzyme is present. The actions of NO in each body site are not necessarily going to be the same in each case, however. For example, NO is formed in the brain, where it may serve as a sort of gaseous intraneuronal communication signal. In the neurons that control activity of the muscles of the gastrointestinal tract, however, NO causes muscular relaxation, and in the walls of blood vessels NO is also a potent muscle relaxant. The latter phenomenon causes blood vessels to open wider, or dilate, and this tends to reduce blood pressure in the body. In fact, it has been suggested that a deficiency in the enzyme required to produce NO may be one cause of hypertension in some individuals. Another area where the vessels respond to the dilatory effects of NO is the human penis. A lack of nitric oxide means little or no dilation of blood vessels, thus no erections!

Anyone who has had, or knows someone with coronary artery disease, is probably aware that nitroglycerin tablets are often used to relieve the symptoms. That is because nitroglycerin, as the first half of its name implies, contains nitrogens and can act as a "donor" of nitric oxide under certain conditions. By releasing NO into the circulation, the coronary artery walls dilate and enlarge, allowing blood to flow more readily to the muscle cells of the heart.

NO has a fleeting existence. It is quickly oxidized in the blood and tissues to NO_2—nitrogen dioxide—an extremely dangerous poison. However, NO_2 is rapidly converted to nitrite (a salt of NO_2) and excreted. Nitric oxide is also formed in air if sufficient energy is available, such as during a lightning storm. Some of that nitric oxide ends up in the soil, where it is used as a source of nitrogen for plant proteins. It can also react with ozone, in which case the result is smog. It is interesting to note that if a second N atom were added to NO instead of a second oxygen atom, the result would be N_2O, nitrous oxide, known as laughing gas. It is truly remarkable that the simple chemical differences of NO, NO_2, and N_2O make the difference between having an erection, being poisoned, or being anesthetized!

[6]

The Framework

Muscles provide the force needed to make bones move. For bones to move, therefore, the muscles must be physically attached to the bones. This is accomplished by means of tough, fibrous tendons that attach at one end to a muscle and at the other end to bone.

The use of a muscle is a wonderfully complex event that requires the cooperative actions of a fair number of molecules. These molecules must interact in a precise way to ensure that the muscle fibers will contract, and thereby provide force. All of this, not surprisingly, requires considerable energy, which is provided by ATP.

Bones, on the other hand, are often considered to be nonliving structures—something for the body to hang on to. But in fact, bone is an active, living tissue that plays important roles in blood cell formation, the immune system, and calcium storage, as well as providing the levers that muscles act on to produce movement. Bone is constantly being re-formed and molded: the skeleton you die with will be completely different from the skeleton in your body right now. Most of the minerals will have been removed and replaced over and over again. In part, this is because the bones represent the major reservoir of calcium for the body. Since maintenance of normal blood levels of calcium is vital for proper brain activity, as well as heart function and muscle contraction, the calcium in bones is constantly being tapped to help regulate blood calcium levels around the normal level.

The basics of how bones are formed and re-formed are not really all that complex. You need some collagen (a protein), some calcium and phosphorus (minerals), a few enzymes, and three major types of cells, all mixed in the right ratios, to get bone to form. With the exception of elasmobranchs (sharks, skates, and rays, which have cartilage instead of bones), all vertebrates have bones that are formed in more or less this same way.

Collagen and Bone

The cells in our bodies need something to support them in space. Without a tough, fibrous network of *connective tissue,* we couldn't maintain our shape. We wouldn't quite take on the appearance of jellyfish, but we'd be a lot less sturdy than we are.

Connective tissue includes bone, tendons, ligaments, and all the fibrous framework that surrounds each individual cell and organ in the body. All of this structural material is made up of protein, primarily collagen. And since much of the body is made up of bone and connective tissue, collagen is among the most abundant proteins in living animals.

As proteins go, collagen is unusual in that only two amino acids (glycine and proline) account for roughly two-thirds of the total number of amino acids in the protein. Prolines are rather bulky, and this fact combined with the electrical attractions between glycine and the other amino acids causes the collagen molecule to assume a characteristic fibril-like shape. In addition, three collagen fibers can intertwine to form a single, tough, helical fiber (figure 8), not unlike the toughness imparted to a rope as thin strands are interwoven with other strands. The intertwining of collagen is possible because glycine is the smallest of all the amino acids, allowing it to fit neatly into the inner grooves of the helix. This permits tight packing of the three strands around one another. Thus, collagen provides a strong, almost inelastic framework that enmeshes the cells and organs of the body. Pound for pound, it is about as strong as steel.

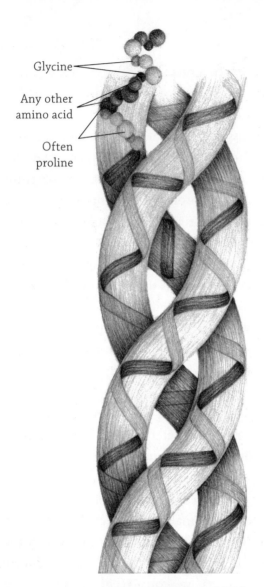

Glycine

Any other
amino acid

Often
proline

Figure 8. COLLAGEN

Each collagen molecule is twisted into a stable helical
shape. Three such molecules combine in an interwoven
pattern to impart great strength to a collagen fiber.

To form bone, collagen fibers must become "mineralized"—that is, they must combine with calcium and phosphorus salts that crystallize out of the watery solution surrounding them. The crystallized salts cement themselves around the collagen fibers, giving bone considerable strength, much the same way that pouring concrete over a meshwork of steel bars reinforces a building. Although collagen is normally insoluble, allowing it to form tendons and bone, upon boiling in water, the three strands come apart and the fibers dissolve. The dissolved fibers do not re-entwine upon cooling, but set as a gel, known as gelatin. The ancient Egyptians knew of this property of collagen and used the gelatinous mixture as the chief ingredient in glue for woodworking.

Vitamin D

One of the most important elements in our diet is calcium. No muscle, including the heart, can contract without calcium. Likewise, the ability of cells to secrete hormones and neurotransmitters depends upon calcium, as does the ability of blood to clot and bones to form. Unfortunately, calcium exists in our digestive tract primarily in its ionized, electrically charged form, and charged particles are normally prevented from crossing cell membranes. Charged particles tend to be stable in watery environments and can cross the lipid-rich cell membrane environment only with assistance.

In the case of calcium, that assistance comes in the form of what is usually referred to as vitamin D.

What we normally call vitamin D begins in our outer skin cells as a precursor of cholesterol (figure 9). It differs from cholesterol by the absence of a single hydrogen atom that is normally attached to the seventh carbon atom in the cholesterol molecule; it is referred to as 7-dehydrocholesterol. This is a relatively stable lipid that remains in the skin until exposed to sunlight. The energy of sunlight is sufficient to break apart one of the "rings" in the structure of 7-dehydrocholesterol, forming a new molecule called vitamin D_3 (also known as cholecalciferol). D_3 is also ingested in the diet, for example, in fish oils. But unless fish oils are a regular part of a person's diet, the sunlight-dependent conversion process will be needed to form sufficient D_3.

D_3 has little or no biological activity. Once produced in the skin, it diffuses into the blood and is carried through the circulation to the liver. The liver contains an enzyme that adds an oxygen and a hydrogen atom (known as a hydroxy group, or OH group) to the number 25 carbon of D_3. Enzymes of the type that transfer OH groups to carbon are widespread in the body and are known as hydroxylases. The end product of this reaction is called 25-hydroxycholecalciferol, which is secreted back into the blood. Yet 25-hydroxycholecalciferol is relatively inactive and must be further "hydroxylated" at the number 1 carbon to make it fully active. This additional reaction takes place not in the liver, but in the kidneys. D_3 must travel, therefore, from the skin to the liver, and then to the kidneys before finally becoming active.

The final hormone, known as 1, 25-dihydroxycholecalciferol, leaves the kidneys and travels to the intestines. There it stimulates the formation of intracellular proteins that help shuttle calcium across the intestinal cell membranes. The calcium then moves through the intestinal cell cytoplasm and out the other side, where it is picked up by blood vessels. In the absence of sufficient sunlight or dietary supplementation with D_3, much of the calcium we ingest will not be absorbed through the intestinal walls and would simply be excreted. We would become calcium-deficient, and the result could be severe. In children, whose bones are still forming and whose long bones have not yet fused, inadequate calcium absorption leads to rickets. This is where the bones fail to mineralize (deposit calcium and phosphate properly) and become easily deformed. In adults, the condition is known as osteomalacia and results in easily fractured bones. Today 1,25-dihydroxycholecalciferol deficiency can result from a combination of any of several factors, including inadequate dietary vitamin D intake, living in northern latitudes where sunlight is limited for half the year, the presence of darkly pigmented skin (melanin, the pigment in skin, partially inhibits one of the steps induced by UV radiation—possibly as a mechanism to prevent overproduction of D_3 in people in regions that are normally exposed to high

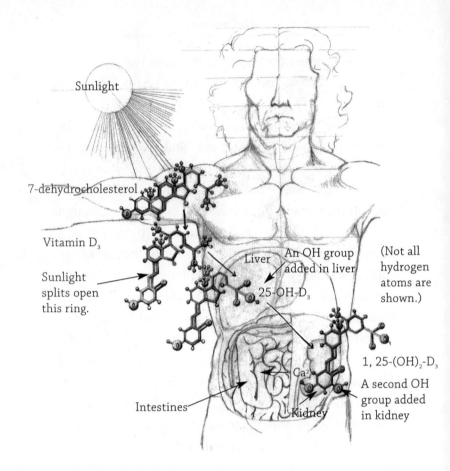

Sunlight

7-dehydrocholesterol

Vitamin D_3

Sunlight
splits open
this ring.

Liver

An OH group
added in liver

25-OH-D_3

(Not all
hydrogen
atoms are
shown.)

Ca^{2+}

Intestines

Kidney

1, 25-(OH)$_2$-D_3

A second OH
group added
in kidney

Figure 9. VITAMIN D

The conversion of precursor molecules in the skin to
the active final compound (1,25-dihydroxycholecal-
ciferol) requires several steps and the participation of
three organs: skin, liver, and kidney.

amounts of UV radiation), or cultural factors that reduce exposure to sunlight (clothing that covers all or most of the body year-round).

Actin and Myosin

There are three types of muscle in the human body. Skeletal muscle, as its name implies, is attached to our bones and is responsible for our ability to move. It is under conscious control. Smooth muscle wraps around our blood vessels, intestines, and other places, and is under involuntary (neuronal) control. The third type, cardiac muscle, is not attached to bone and is not under voluntary control. Cardiac muscle cells are also electrically coupled to each other in a unique way that allows the heart to behave almost as if it were one giant muscle cell. A last major difference among muscle types is that some, like cardiac muscle, have built-in pacemakers that cause them to contract at regular intervals, while others are relatively dormant until called upon for action.

Within a single muscle are numerous muscle cells, called fibers. Each fiber, in turn, is composed of the usual cellular features: nucleus, mitochondria, and so on. But unlike other cells in the body, muscle fibers are literally loaded with parallel arrays of two major proteins called actin and myosin (these are also collectively called the filaments of a muscle). Actin is a polymer, or chain, of smaller molecules, which link together to form a strand-like fiber. Closely twined around each actin filament are two other molecules, called troponin and tropomysin. And stacked between actin fibers are the golf club–shaped myosin proteins.

Thus, within a muscle fiber are parallel stacks of alternating rows of actin and myosin. The myosin filaments have a long, doubly-wrapped tail segment, and two "head" regions that bend up or down from the tail at an angle. When a muscle cell is stimulated, the membranes of the muscle cells become leaky for calcium ions. The calcium outside the cell rushes into the cell cytoplasm, where it encounters troponin. Troponin can bind to calcium, and when it does the precise arrangement of actin, troponin, and tropomysin molecules becomes deformed. In effect, the deformed troponin pushes the tropomysin aside. This exposes a region of the actin molecule that was previously hidden. The exposed region binds to the heads of the myosin molecules. Within the head region is an enzyme that metabolizes ATP, which you will remember is the major form of stored chemical energy in all cells. Enzymatic metabolism of ATP in the muscle cell releases the pent-up energy from ATP. When this energy is released, it can be used to move the myosin heads in one direction, dragging the actin fibers with them.

The sliding of actin and myosin fibers back and forth over each other causes the shortening (contraction) of a muscle. Contraction occurs because myosin molecules within a muscle fiber are laid down back to back. In other words, the heads of one group of stacked myosin molecules are oriented toward one end of the muscle, while the heads of the adjacent stack are oriented the other way. So when both groups of myosins pull on the actin, the actin filaments at one end of the muscle fiber are pulled closer to the actin filaments at the other end, and the total length of the muscle shortens. As it shortens, the muscle pulls on a bone and causes it to move, or, in the case of the heart, the contraction causes a heartbeat.

Another interesting feature of muscles is that they can be stimulated to enlarge when used more frequently. But new muscle cells are not usually grown; rather, additional muscle fibers are formed within each muscle cell. Certain hormones, like testosterone, are also capable of causing muscle fiber growth. Likewise, if muscles are not used for long periods, as in long space missions, they tend to atrophy.

Acetylcholine

In order to get muscles to contract, a signal of some sort must be sent to the muscle cells so that contraction occurs at the appropriate time and with the appropriate strength. It would hardly do to have your bicep contract too forcefully while trying to bring a forkful of food to your mouth.

The signal received by all muscle cells involved in movement is from a small molecule called acetylcholine that is released by nerve cells that come into contact with muscle cells. Acetylcholine is composed of one molecule of acetic acid and one of choline. These two substances are chemically combined within the nerve cells that cause movement, the motor neurons. Without the input from a motor neuron, a muscle would not be able to contract. In fact, in the complete absence of neuronal stimulation, the muscle would eventually atrophy due to disuse.

Here's how acetylcholine works. A motor neuron produces acetic acid as a breakdown product of metabolism and combines it with the small choline molecule by means of an enzyme. The newly formed acetylcholine is packaged within the nerve ending in small, membrane-bound vesicles. When a signal is sent from the brain (for example, "Lift the fork"), the vesicles dump their contents to the space immediately outside the nerve ending and immediately above the muscle cell membrane. This space is known as a synapse. The acetylcholine drifts through the synapse, reaches the muscle cell, and activates a specific membrane receptor molecule (again, a protein). Once the receptor is activated, the muscle cell is now cued to the fact that the brain wants the muscle to move. Contraction occurs by the mechanism outlined in the previous section. Depending on the strength of the message from the brain, the contraction is either great or small ("Lift the fork slowly, so you don't stab yourself in the mouth"). The strength of the contraction depends entirely upon the amount of acetylcholine released.

It's important that the acetylcholine is removed from the synapse once its job is done, or else it would cause the muscle to become permanently contracted. Removing acetylcholine is the job of an enzyme located in the synapse; it destroys the acetylcholine and converts it back into acetate and choline. Acetate and choline can then be recycled back into the nerve cell and re-formed yet again into new acetylcholine. If the enzyme, known as acetylcholinesterase, fails to function properly, acetylcholine levels will build up within the synapse and the muscle will be overstimulated. The result of overstimulation by acetylcholine can be devastating. Nerve gas, for example, works by blocking the action of acetylcholinesterase; one result of exposure to nerve gas is spastic, uncontrolled contractions of respiratory muscles leading to suffocation.

The actions of acetylcholine are not restricted to the brain-muscle interface. In fact, acetylcholine is found throughout the brain. It seems to have diverse actions, ranging from control of hormonal secretions to the psychological effects of cigarette smoking, to a possible involvement in Alzheimer's disease. Some of the common drugs used in medicine today are based on their ability to either mimic or inhibit the action of acetylcholine. Atropine, for example, is a drug derived from a plant known as deadly nightshade. Atropine blocks the ability of acetylcholine to bind to acetylcholine receptors on certain muscle cells. Thus, in the presence of atropine, acetylcholine cannot exert its actions. Atropine is used to dilate the pupils during eye exams, by paralyzing the muscles of the iris. Atropine has also been used to help open constricted airways in asthma, by relaxing the smooth muscles that surround the bronchioles. Atropine, in fact, is also the antidote for people who have been exposed to nerve gas. Atropine gets its old name, belladonna (pretty woman), from the practice of the ancient Italians, who considered dilated pupils in women a sign of beauty. Anyone who has ever had his or her pupils dilated during an eye exam will undoubtedly wonder why anyone would willingly do this to themselves. There must have been a lot of squinting back then.

[7]

Sex and Hormones

Most of the molecules we have encountered so far make up a significant portion of the body's show. It is up to the hormones to make sure that the show takes place. Hormones control sex drive, fertility, blood pressure, sugar balance, salt and water levels, bone formation, growth, development, mood, appetite, and heat production, to name just some of their functions. Remarkably, all of these varied functions are carried out by just three chemical categories of hormones: those whose structure is derived from cholesterol, those that are proteins, and those that are derived from an amino acid. These differences in structure have profound effects on the actions of these hormones, their ability to dissolve in the blood, and their ability to enter the interior of cells.

Exactly what is a hormone? Most people are quite familiar with the term and could no doubt name one or more (typically the sex hormones estrogen, testosterone, and progesterone). It might be harder to actually define what a hormone is, though. Simply stated, a hormone is any chemical that is produced in one part of the body, secreted into the bloodstream, and transported to another part of the body to carry out some biological action. In other words, there are glands scattered throughout the body (endocrine glands) that make, store, and secrete hormones into the blood. Once in the blood, the hormone can travel to any cell anywhere in the body. If that cell has on its surface (or in some cases, within its interior) a protein receptor that specifically recognizes that hormone, the cell will respond to the hormone. Thus, testosterone will bind its receptors in muscle cells and initiate muscle growth. It will not initiate growth of other structures, such as the muscles of the iris, because the necessary receptors are not located there.

In recent years, an extraordinary profusion of "new" hormones has come to light. Many organs in the body not previously recognized as having hormonal functions have now been recognized as such. The heart, for example, is now known to make a hormone called atrial natriuretic peptide, which causes the kidney to increase the excretion of sodium (natrium) and water (diuresis) during times when the fluid level in the body rises. Likewise, the pineal gland (in the brain), the skin, the liver, the kidneys, and even adipocytes (fat cells) are, in addition to their other functions, endocrine organs. Without certain hormones, like cortisol and insulin, we cannot survive. Without others, like thyroid hormone, estrogen, and testosterone, we can survive but the quality of life is adversely affected. Indeed, hormones are so critical to our daily lives that it is impossible to overstate the importance of the integrated network of hormonal signaling that must be in place from the earliest stages of embryonic life right up to the day we die.

Cortisol

Cortisol is a hormone that belongs to a general class of hormones known as steroids. All steroid hormones are lipids and are formed using cholesterol as a template. The chemical differences between different steroids are sometimes very slight, but their functions are vastly different. The addition of a single oxygen and hydrogen atom can create two different hormones like cortisol and cortisone. In this case, the latter contains what's known as a ketone group (a carbon atom that shares not one but two electrons with oxygen) at a key position, where cortisol contains an hydroxyl group (a carbon atom sharing one of its electrons with oxygen, which itself shares another one with a hydrogen atom). Cortisol is sometimes confused with cortisone, but cortisone is relatively inactive and cortisol is extremely potent. In pharmacies, dilute creams and lotions of cortisol are sold over-the-counter as hydrocortisone. This unnecessarily obtuse term merely means cortisone with a "hydro" (hydroxyl) group instead of a ketone; in other words, cortisol. Hydrocortisone and cortisol are one and the same.

Cortisol is made in the outer part of the adrenal glands known as the adrenal cortex. (It is the inner part of the adrenal, known as the adrenal medulla, that secretes epinephrine, also known as adrenaline.) Cortisol is absolutely essential for life, a phenomenon first predicted by the British physician Thomas Addison in 1865. It is one of a group of hormones known as glucocorticoids, because one of cortisol's key functions is to maintain blood levels of glucose within normal limits. It does this partly by responding to the body's need for glucose by breaking down bodily tissues. The breakdown products of muscle, bone, immune tissues, and fat can all be used as substrates for the liver to convert into glucose. Because of its ability to break down body tissue, cortisol is a *catabolic* steroid, not to be confused with other steroids like testosterone, which build up tissue (such as muscle), and are called *anabolic.*

Another function attributed to cortisol is its apparent "check" on the activity of the immune system, including inflammation. High levels of cortisol inhibit immune responses, an action that doctors exploit when trying to suppress rejection of transplanted organs. For similar reasons, cortisol creams are effective in treating certain inflammatory diseases and skin conditions (poison ivy and eczema are two examples). In normal health, the anti-immune actions of cortisol are believed to minimize the possibility of our immune system attacking our own cells in an overzealous attempt to clear the body of potentially dangerous cells and compounds.

Among its many other actions, cortisol also helps maintain blood pressure within normal limits, partly by enhancing the action of other hormones like epinephrine that are potent blood pressure–raising molecules. Last but not least, cortisol plays crucial roles in fetal development, particularly in the formation of the brain and lungs (recall that it is a chief factor in the production of surfactant).

The effects of too little or too much cortisol can be predicted from the physiological actions of cortisol. Too little cortisol leads to hypoglycemia, low blood pressure, and an overactive immune system. Known as adrenal insufficiency (Addison's disease is a type of adrenal insufficiency), it can be fatal if not treated properly. Too much cortisol, on the other hand, leads to Cushing's syndrome (after the American physiologist and surgeon Harvey Cushing, its discoverer), characterized by elevated blood sugar, hypertension, immunosuppression, and increased catabolic activity, resulting in bone and muscle loss.

The POMC Gene

The so-called stress response in humans, and in all mammals for that matter, starts with several key hormones being rapidly released into the blood. Already discussed is cortisol, secreted from the adrenal glands. Two other important stress hormones are produced in the pituitary, and these illustrate the remarkable efficiency with which the human genome operates.

In the portion of the pituitary gland known as the anterior pituitary (because of its location toward the front—or anterior—part of the head), a single gene is activated during times of physical or psychological stress. This gene, when processed by the machinery of the pituitary cell, results in the production of not one but several protein hormones, all of which are involved in some way in the response to stress. Rather than evolve multiple genes, each coding for a particular protein, nature has apparently decided that it is more efficient to have a single long gene whose protein product can be sequentially chopped up into smaller and smaller proteins. One gene, many proteins.

The gene in this case is known as the POMC gene, for pro-opio-melano-corticotropin. *Pro* indicates that this is a gene with many possible products; *opio, melano,* and *corticotropin* stand for three of the major products. The first of these is a molecule known as beta-endorphin. This molecule is a natural painkiller, in the same chemical class as opiates (hence, *opio*). It is released during stress, presumably to combat pain; think of people who report not feeling the pain of a major injury until much later, when they are out of harm's way. It is probably also the hormone responsible for the "runner's high," experienced by people who exercise at an intense level for long periods of time (the brain interprets such extreme exercise as a "stress" to the body). The second molecule is a hormone called melanocyte-stimulating hormone, a hormone responsible for seasonal changes in coat color in some mammals, but whose physiological role in humans is still debated. The third major product, adrenocorticotropic hormone, or ACTH, is the pituitary hormone that signals the adrenal glands to start producing cortisol.

This process whereby a single gene produces a protein that is spliced up into multiple products is not unique to the pituitary gland. In fact, it's a fairly common feature of the genome, since it's a good way to get more bang for your buck each time a gene is turned on. Some *potential* hormones have even been identified this way, though no function has yet been attributed to them. For example, there are smaller bits of the POMC molecule that are clipped off from the "parent" molecule within the pituitary. We know their chemical structure, and we know how and when they are produced, but we don't yet know what function these small putative hormones may have. It is a mystery worth investigating, however, since nature rarely creates things for no reason.

Epinephrine (Adrenaline)

The terms epinephrine (Greek) and adrenaline (Latin) mean the same thing, namely, originating "above" or "toward" the "kidney." The Latin *adrenaline* is derived from "adrenal" glands. Nowadays, for no particular reason, the term epinephrine is used in the United States, but adrenaline is used in most of the rest of the world. Nonetheless, we still refer to the "adrenal" glands here and not to "epinephros" glands. The confusing nomenclature notwithstanding, epinephrine is a vital hormone that, like cortisol, ACTH, and endorphin, constitutes part of the defense mechanism the body employs to combat stressful challenges.

The adrenal gland is actually two glands in one; the outer layer, called the cortex, makes steroid hormones like cortisol, and the inner layer, the medulla, makes epinephrine. The medulla is actually an extension of the nervous system that has migrated into the adrenal gland. As such, the response time between the onset of stress (pain, injury, drop in blood pressure, lack of oxygen, etc.) and the appearance of epinephrine in the blood is exceedingly fast, on the order of seconds. Epinephrine acts to promote improved heart function, better airway ventilation (by dilating the passageways in the lungs), and increased availability of fuels in the blood (glucose and fatty acids). Each of these functions is potentially critical in surviving life-threatening stresses and they are major features of the so-called fight-or-flight response to stress.

For all its importance, the chemical structure of epinephrine is disarmingly simple. It begins life as a simple amino acid, tyrosine, which is obtained through the diet. Tyrosine is converted within the adrenal medulla by enzymes to an intermediate called L-dopa, then to a second intermediate, dopamine, and then to norepinephrine (noradrenaline). The latter is acted upon by yet another enzyme to form epinephrine. The final step, whereby norepinephrine is converted to epinephrine, is stimulated by cortisol, which activates the final enzyme in the chain of reactions. This need for cortisol may be why the adrenal medulla forms within the adrenal cortex during fetal development: the close proximity ensures that the medulla will always be bathed in the high levels of cortisol produced by the neighboring cortex cells. Incidentally, the intermediate dopamine is also made in the brain, where it functions as a neurotransmitter. In dopamine-producting brain cells, the final enzymes needed to continue the reactions all the way through to epinephrine are missing, which ensures that the reactions stop at dopamine. This kind of sequential processing of a single amino acid into more than one possible active molecule is also seen in the pineal gland, where tryptophan can be converted to either serotonin or melatonin (see chapter 8).

Of clinical importance is the fact that cocaine and

amphetamines act, in part, by mimicking or enhancing the actions of epinephrine and norepinephrine. That is why ingestion of these drugs causes the heart to race and blood pressure to increase.

Testosterone and Estrogen

Chemically speaking, it's remarkable how little difference exists between a man and a woman, or any male and female animal. Genetically, as is well known, there is one major difference that resides in the so-called sex chromosomes. Females have two X chromosomes, while males have one X and one Y chromosome. But it is the products of the different genes on the Y chromosome that make an embryo develop male characteristics. In the absence of proteins directed by the Y chromosome, an embryo will become a female. In essence, femaleness is the default mode.

A key difference that arises early on in this process of female/male differentiation is the formation of either ovaries or testes. This is critical for the rest of the developmental process, because the products from these glands are responsible for the continued formation of the internal and external genitalia associated with males and females. If, for example, a female embryo or fetus were to be bombarded with male hormones, she would begin showing some of the physical characteristics of maleness. This can happen in rare situations where the gonads or adrenals fail to function properly during development. For instance, a female (XX) fetus with an adrenal disorder called congenital adrenal hyperplasia may have unusually high levels of male hormones in her blood. This results in the formation of external genitalia that may show masculine appearances. These same hormones are also the critical factors driving gender-appropriate development at puberty.

The two key hormones involved in these processes are testosterone, a member of the androgen class of hormones, and estrogen. Both are steroids, like their chemical cousin cortisol. That is, both are lipids and are derived from cholesterol, which is one of the reasons why cholesterol is an important and necessary component of living tissue. Some steroids, like cortisol, are catabolic; they break down tissue to supply the liver with potential fuel sources. Other steroids, like testosterone, are strongly anabolic; they build up tissue (for example, muscle). Still others, like estrogen, are neither greatly catabolic nor anabolic (except in certain regions where they may promote growth, as in the uterus).

Returning to the chemical difference between men and women, let's look at the structures of these two steroid molecules (figure 10). During the conversions from cholesterol to testosterone, several intermediates are formed along the way, though some of these are believed to be of little physiological consequence. One exception, however, is progesterone, which is manufactured by the gonads and also by the adrenals of both sexes, and which is an important hormone in

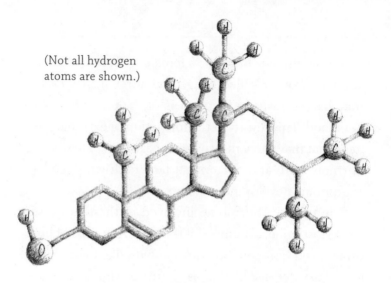

(Not all hydrogen atoms are shown.)

Figure 10. CHOLESTEROL

All of the steroid hormones are made from cholesterol, including the three major reproductive steroid hormones shown in this figure.

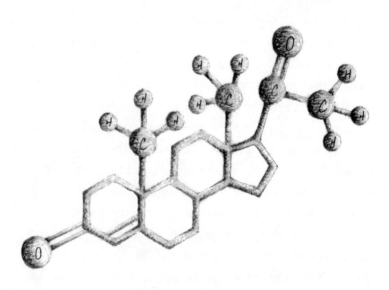

PROGESTERONE

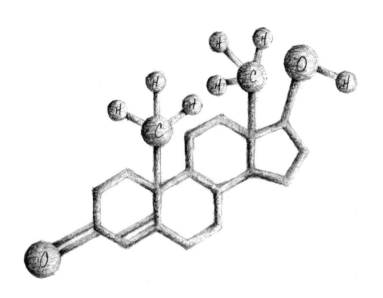

TESTOSTERONE

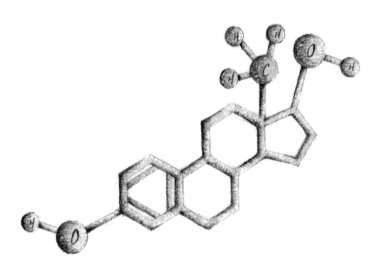

ESTRADIOL

maintenance of pregnancy (progesterone = PROmotion of GESTation). In addition, progesterone has a slight effect on body temperature elevation and increases respiration rate in pregnant women. Thus, pregnant women not only eat for two, they breathe for two!

By the time cholesterol has been converted all the way to testosterone, the molecule looks vaguely similar to cholesterol but has been modified quite a bit. The jump from testosterone to estrogen, though, is just one step. An enzyme modifies the bond between a carbon atom and an oxygen atom in one of the "rings" of testosterone, and chops off one of the nineteen carbon atoms in testosterone in the process. Thus, a single (albeit complex) set of reactions converts testosterone to estrogen, and accounts for the physical differences we observe between men and women: nineteen carbons, and you can grow a beard, eighteen carbons and you can give birth. It is rather poetic to realize that the presence or absence of this single chemical reaction accounts for much of the joy in human life!

Technically speaking, estrogen is the name given to a family of related compounds. The name is derived from the Greek *estrus,* or "intense desire." Steroid hormones that promote estrus (mating) behavior in seasonal animals are called estrogens. In humans, the primary estrogen is called estradiol.

In nature, there exists a multitude of naturally occurring compounds that resemble estradiol in structure. Most of these occur in plants, which may then be eaten by animals. Thus, the feed given to farm animals must be monitored to minimize exposure to these "estrogenic" compounds. Still other molecules, quite different in structure from estradiol, are able to cause estrogen-like effects in humans and other animals. These chemicals, often called xenobiotics, can mimic the actions of estradiol because they are able to bind to the same cellular receptor molecule that mediates estradiol's activity. Any compound that activates a receptor will induce a biological response, even if that compound is not the natural one to which the receptor is normally exposed. The growing presence in our environment of xenobiotics like the heavy metal cadmium is of considerable concern because of the possible effects on reproductive fertility in people exposed to chronic, aberrant estradiol activity. A final class of estrogen-like chemicals are those that are synthesized in laboratories for therapeutic uses. Tamoxifen is a nonsteroidal drug that can nonetheless bind to estradiol's receptor molecule. Depending on the dosage, it can mimic or block the ability of the body's own estradiol to do its job. It is used for a variety of purposes, notably for the treatment of breast cancers whose growth has been shown to be stimulated by estradiol.

The reason men make testosterone and little estrogen, and women make primarily estrogen but little testosterone, is that the ovaries contain the enzyme needed for the conversion of testosterone to estradiol, but the testes do not. It's really that simple. The Y chromosome directs the formation of testes, which make only minimal amounts of estrogen; therefore the process of steroid formation in men pretty much stops at testosterone, which promotes further male development throughout the rest of life.

The adrenal glands also produce some testosterone and also some estrogen, in both sexes. Men and women thus have both hormones, but in reverse ratios. When that ratio changes, even in adulthood, the physical characteristics of an individual can change. The changes might not be as dramatic as they would have been had they occurred during embryonic development, but they are still rather impressive. For instance, the ovaries of women after menopause begin making less and less estrogen. This reduces the estrogen-testosterone ratio in women, and signs of masculine characteristics may sometimes appear. For example, some of the protective actions of estrogen—such as its beneficial effects on cardiovascular function—begin to decline, and post-menopausal women show a similar risk of atherosclerosis as their testosterone-laden male counterparts. As men age, testosterone levels slowly decline and estrogen levels rise a bit, but the changes are nowhere near as dramatic as in women.

Thyroid Hormone

The thyroid gland sits just below the larynx in the throat. Because of this positioning, it is not difficult for a physician to palpate someone's thryoid gland for signs of overgrowth or underdevelopment. An overgrown thyroid gland is known as a goiter and can grow to the size of a cantaloupe. Some goiters get large enough to impede a person's range of head movement and may even hinder air movement through the trachea.

The hormone released from the thyroid gland is called thyroid hormone, but really exists in two related forms. The difference between these forms is simply the number of iodine molecules associated with them: tri-iodothyronine (T3) has three iodine molecules within its structure, while thyroxine (T4) has four. Of the two, T3 is the more potent hormone.

Iodine is what gives thyroid hormone its unique characteristics. T3 and T4 are the only hormones in the body that require incorporation of iodine. Roughly 98 percent of all the body's iodine is concentrated in the thyroid. Thus, proper formation of T3 and T4 depends upon an adequate intake of dietary iodine (about 150 micrograms a day).

The iodine in our diet is transported into the cells of the thyroid gland, where the iodine is attached to a carbon "ring" structure within the amino acid tyrosine. This is the same amino acid that forms the backbone for dopamine and epinephrine. The tyrosines within the thyroid gland cells are part of the structure of a large protein called thyroglobulin (a globular-shaped protein within the thyroid). There is roughly a three-month supply of thyroglobulin and its tyrosines within a normal, healthy thyroid gland. No other endocrine gland has such a storage capability. Undoubtedly, this evolved in the thyroid to compensate for the rareness of iodine in the diet.

The effects of thyroid hormone are manifold. T3 increases metabolism, which raises heat production in the body. It also aids in normal menstrual function, brain activity, reflexes, heart function, and growth and development. In adults, a deficiency in T3 leads to a state known as hypothyroidism, characterized by weight gain (due to decreased metabolism), sluggish reflexes, tiredness, and cold intolerance. Hypothyroidism is sometimes the result of inadequate iodine intake, but more often is due to autoimmune destruction of the thyroid. Iodine is not common in food, but nowadays table salt is iodized specifically to provide a source of iodine. (About 1 molecule in 10,000 of each NaCl in table salt is replaced with a molecule of NaI, sodium iodide.) Despite the availability of iodized salt, it is somewhat distressing to note that roughly one in six people around the world are still at risk for developing iodine-deficient hypothyroidism.

When dietary iodine intake is deficient during pregnancy, the consequences for the developing fetus can be catastrophic. Insufficient iodine in the mother's blood means insufficient iodine crossing the placenta and reaching the blood of the fetus. The brain of the growing fetus is critically dependent upon normal levels of thyroid hormone. Without iodine, less fetal thyroid hormone is made (maternal thyroid hormone cannot compensate since it does not get across the placenta), and the result is a profound form of mental retardation known as cretinism.

Excess thyroid hormone production, called hyperthyroidism, results from a number of causes, notably thyroid tumors producing uncontrolled amounts of the hormone, and Graves' disease. Graves' is an autoimmune disease in which the body mistakenly recognizes a protein on the surface of thyroid gland cells as being "foreign." Antibodies are made against the supposedly foreign protein. However, when the antibodies combine with the protein, it causes a chronic activation of the thyroid. The gland grows in size and forms a goiter, while making excessive amounts of T3. The result is hyperactivity, nervousness, restlessness, weight loss, and heat intolerance due to the high metabolism. It is treated by taking advantage of the fact that the thyroid gland sequesters almost all of the iodine ingested in the diet. By treating a patient with Graves' disease with radioactive iodine, the iodine ends up in the thyroid and the radiation kills off most of the goiter. It is in essence a targeted form of radiation therapy. Radioactive iodine is also a product of nuclear fission reactions and appears in the air after an atomic explosion. When inhaled, the radioactive iodine enters the blood and is taken up by the thyroid. Not surprisingly, people exposed to radioactive fallout have a higher than normal likelihood of future thyroid disease and thyroid cancer. People treated with radioactive iodine for Graves' disease, however, generally do not develop symptoms of radiation sickness or thyroid cancer, because the treatment is brief and less intense than exposure to radioactive fallout.

[8]

The Brain: Perception and Behavior

The brain is composed primarily of two types of cells. The first, known as glial cells, are support cells for the other type, known as neurons, or nerve cells. Although glial cells have important functions in maintaining the metabolic microenvironment around neurons, all of the cell-to-cell signaling in the brain occurs via neurons. Thus, this chapter focuses exclusively on neurons and the chemicals within them.

Regions of the brain must communicate not only with other brain regions but also with structures that are outside the brain, such as muscles or the heart. Communication between the brain and these other structures also relies upon neurons. Electricity, in the form of flowing ions—primarily sodium and potassium—

moves through the neurons to reach the muscles, heart, and other structures. That's one reason why a stable level of these ions is so vital to survival. Within the human brain there are about a trillion neurons. Each one of them has the capacity to receive information from dozens to tens of thousands of other neurons, and they can process all that information simultaneously. It is the incredible permutations of these numbers that makes the brain the fantastically complex structure it is.

The information transmitted through these communication pathways is stored in neuronal endings in the form of "neurotransmitters." There are many such neurotransmitter molecules, and most are small, simple compounds like acetycholine or norepinephrine. Some neurotransmitters are excitatory: when released from the ending of one neuron to the beginning of another, the second neuron becomes electrically stimulated, too. Other neurotransmitters are inhibitory: they slow down other neurons. So, you might say that the brain always operates with one foot on the accelerator and the other on the brake. During periods of alertness, the ratio of transmitter activity shifts in favor of the accelerator, while in periods of relaxation or sleep, the brake is applied more than the accelerator. The balance between these can shift almost instantaneously. Imagine that you are dozing off in a comfortable chair (riding the brake), and suddenly someone lights a firecracker outside your window (the brake comes off and the accelerator slams down).

Understanding the physiology of neurotransmitters (what they do and how they do it) and their pharmacology (how we might initiate or manipulate the actions of neurotransmitters with therapeutic drugs) is a major feature of modern neurological medicine. In this chapter, we'll look at three major neurotransmitters whose chemistry and physiology is especially well understood, and whose role in certain diseases is also established. We'll also look at other key molecules within the brain that are not neurotransmitters but are representative of the numerous brain proteins responsible for everything from visual light reception to learning and memory.

Dopamine

For such a relatively small molecule, dopamine's importance in normal brain activity is enormous. The process that creates this neurotransmitter is surprisingly simple. It only takes two chemical reactions, each catalyzed by enzymes, to convert the ordinary amino acid tyrosine (derived from food) into the neurotransmitter dopamine. In the first step, an intermediate of little or no biological activity is generated, L-dopa, which, in turn, is converted by another enzyme into dopamine.

Since there are many different classes of receptors for dopamine scattered throughout the brain, dopamine can exert a wide array of effects. Each receptor may be linked with a different behavior. For example, an excess of the so-called type 2 dopamine receptor is associated with the symptoms of schizophrenia (excess receptors means higher sensitivity to dopamine). Likewise, excess production of dopamine can overstimulate receptors, leading to symptoms similar to those of having excess receptors.

In Parkinson's disease, where normal control of motor function is lost, there is a degeneration of the dopamine-producing cells in one part of the brain. Thus, one treatment for Parkinson's is to restore brain levels of dopamine, if possible.

Unfortunately, oral administration of dopamine is not particularly effective because dopamine cannot penetrate the protective sheaths that normally screen the brain from other compounds in the blood—the so-called blood-brain barrier. L-dopa can get into the brain, however, and once there can be converted to dopamine. Nowadays, synthetic L-dopa is used to treat Parkinson's, often with good results. But it is sometimes hard to adjust the dosage of L-dopa precisely, and people taking L-dopa have been known to hallucinate and show other schizoid behaviors because they have been overcompensated for the loss of dopamine. If we continue the logic that excess dopamine in the brain can cause schizoid symptoms, then it should be possible to treat schizophrenia with drugs that block the ability of the dopamine in a person's brain to bind to the dopamine receptors. If dopamine can't bind its receptors, it cannot cause the schizoid symptoms. Such drugs do exist and are often very effective at alleviating psychotic episodes. In some cases, however, the drugs prevent dopamine's actions so well that they induce the symptoms of dopamine deficiency, similar to Parkinson's disease! A share of the Nobel Prize for Physiology and Medicine in 2000 was awarded to Dr. Arvid Carlsson for his discovery of this important neurotransmitter and its relationship to Parkinson's disease and schizophrenia.

Sometimes the illicit use of amphetamines may also result in a schizoid pattern of behavior, known as amphetamine psychosis. That's because within the brain amphetamines act to increase the amount of dopamine released from nerve endings. Thus, drugs as different as amphetamines and L-dopa can have the same result on behavior; the common link is that they alter dopamine content in the brain.

Although dopamine's actions are generally related to the ability of the brain to control movement and regulate behavior, dopamine can also act as a hormone. Released from the neurons in a structure at the base of the brain—the hypothalamus—dopamine travels through blood vessels to the nearby pituitary gland. There, dopamine acts as a chronic inhibitor of the hormone known as prolactin (which gets its name because it PROmotes LACTation). After birth, if a woman is nursing, her hypothalamus secretes less dopamine, which allows prolactin levels to increase in her blood. This will stimulate milk production for as long as nursing continues, even for several years.

Serotonin and Melatonin

Although it was discovered originally in platelets and in the gastrointestinal tract, serotonin has received its greatest fame as a neurotransmitter widely present in the brain. Like most neurotransmitters, it has a simple structure derived from an amino acid, but unlike dopamine and epinephrine, it is produced from trypto-phan, not tyrosine. Only two chemical transformations must take place to convert tryptophan from the diet into serotonin; both are regulated by enzymes, so only those nerve cells that contain these two enzymes will convert tryptophan into serotonin.

Serotonin has been associated with aggression, body weight control (an appetite suppressant), hormone secretion, sleep/wake cycles, and mood. The actions of serotonin depend on the numerous types of receptors found scattered throughout the regions of the brain. Each type of receptor seems to be linked to a different behavior or function. Because of the wide array of func-tions attributed to serotonin and its receptors, it is not surprising that drugs that act by causing serotonin to accumulate in synapses have numerous clinical effects. Drugs like Prozac, Paxil, and Zoloft are known as serotonin-specific reuptake inhibitors (SSRIs) because they block the ability of nerve cells to recapture excess unused serotonin from a synapse. To understand what that means, you need only consider that each time a neuron fires an impulse, it releases an abundance of the neurotransmitter into the synapse with a neighboring neuron. This excess of neurotransmitter ensures that the neighboring cell will be stimulated. It is, however, somewhat inefficient to release so much more transmitter than is actually needed, and so some of it is taken up, or recaptured, by the first neuron and reused in the future.

Now imagine that for some reason the serotonin-containing neurons were not capable of producing sufficient serotonin. Maybe there is a defect in one of the enzymes needed to act on tryptophan, for example. Whatever the cause, if less serotonin is secreted into each synapse, then the effects of serotonin will naturally be reduced. One consequence of this appears to be a clinical depression. Thus, any drug like an SSRI that can block reuptake of serotonin back into the original neuron will in effect allow the serotonin to stick around in the synapse longer than usual. This will give each serotonin molecule a greater chance of reaching the second neuron and stimulating it. SSRI drugs often do work, but side effects can be common. Since excess serotonin is linked with appetite suppression and aggression, for example, depressed patients who are taking very high doses of serotonin reuptake inhibitors, or who are uncharacteristically sensitive to such drugs, may lose weight, develop insomnia due to disruption of sleep/wake patterns, and possibly (rarely) have increased aggressive tendencies.

With just two more chemical modifications, serotonin is converted into melatonin, an entirely different molecule with completely different functions. This occurs primarily in a gland known as the pineal gland. The pineal is centrally located within the brain, and because of this was considered by the ancients to be the seat of the soul. In fact, it is the seat of the enzymes that convert a neurotransmitter into a hormone. Melatonin is released into the blood and acts as a hormone whose precise functions are still being debated. Nonetheless, it appears to be closely linked to both reproduction and the daily, or circadian, rhythmicity of many body functions.

Interestingly, the activity of the enzymes involved in the formation of melatonin are indirectly inhibited by light entering the eyes. A signal is sent from the eyes to the pineal by a rather circuitous route, shutting off melatonin production. This is apparently the way that the pineal gland tells night from day and thus links the day/night cycle with hormone production and daily body rhythms.

Gamma Aminobutyric Acid (GABA)

In the analogy of the brain operating with both the accelerator and brake applied simultaneously, norepinephrine is a chief component of the accelerator. The brake assumes the form of another neurotransmitter called gamma aminobutyric acid, or GABA for short. In a car, pressure on the brake causes slowing of four different wheels simultaneously. In the brain, GABA can exert its inhibition over millions of brain cells all at once. There is hardly any place in the brain where these brake pedals are not present.

Were GABA not present in its usual amounts within the brain, many neurons would be hyper-excitable (remember, the accelerator is always on). Electrical impulses would be sent from neuron to neuron in a haphazard, uncoordinated way. Such a condition sometimes exists in animals, including people, and may be part of the chemical basis of some forms of epilepsy.

The way in which GABA exerts its slowing-down actions is intriguing. Like all such molecules, it must first bind with a protein receptor on the surface of whatever cell it is acting upon. Those receptors are called the GABA receptors, and they come in several varieties. The most interesting of these is the so-called GABA-A receptor. When GABA binds to one of these receptors, it changes the three-dimensional shape of the receptor, causing it to open a pore in the cell surface membrane. This pore, or channel, allows molecules of negatively charged chloride ions to pass from outside the neuron to inside. This makes the neuron less electrically excitable (the more negative charges in a neuron, the less likely it is to fire off an impulse from one neuron to another).

What's odd about the GABA-A receptor is that it is composed of five separate proteins, or subunits, that come together as a single molecule, in a way that permits them to bind GABA and form the chloride pore.

However, there are at least eight possible protein subunits that can be used to form the final, five-subunit receptor. Thus, there really is no single GABA-A receptor at all. Instead, there is a hodgepodge of related receptors with different combinations of the eight possible subunits scattered all over the brain. Each version of the receptor has different abilities to bind GABA and, as it turns out, a host of other compounds. Presumably, this was a way for the brain to make better use of GABA, by having multiple receptor subtypes each of which can bind GABA but may be linked to different functions within the brain.

Drugs used to treat anxiety, like the widely prescribed benzodiazepines (for example, Valium, Xanax), act by binding to a region of the GABA-A receptor. When they do so, they facilitate the ability of GABA itself to activate the receptor. Thus, these drugs help slow the brain down—they have a tranquilizing effect. Ethanol, the active component of alcoholic drinks, also binds to this receptor, and it, too, is a depressant. There is even a site on the receptor that can bind drugs that block GABA's action; such drugs cause convulsions.

Why binding sites exist on the GABA-A receptor for molecules like benzodiazepines is uncertain. Obviously, these didn't evolve so that twenty-first-century people could invent drugs to treat anxiety. There must be natural compounds present within the brain that exert modulatory effects on the GABA-A receptor, although these have yet to be definitively identified.

Opsins

In addition to controlling behavior and thought, the brain must be able to interpret changes in the environment. Sensory cues are detected by specialized sensory receptor cells in different parts of the body (eyes, ears, nose, skin, tongue). Electrical signals are transmitted from the sensory cells to the brain to be decoded and analyzed. One such sensory cue that is especially important for daylight-active animals like ourselves is light.

Light travels in waves of energy called photons. When a photon of light strikes the retina at the back of the eye, it triggers an electrical event in special light-sensitive cells called rods and cones (named for their characteristic shapes). The link between the capture of a photon on a rod or cone and the subsequent electrical signal that travels to the brain is a purely biochemical one. Just as a hormone must bind a receptor on a cell membrane in order to activate the cell, so too a photon must "bind" a receptor of its own.

Remarkably, the receptor for photons resembles the many types of hormone and other membrane receptors found in the human body, even though those other receptors evolved to detect chemical signals, not energy. Yet in the case of the rod and cone, the receptor, called an opsin, doesn't physically bind to anything, since photons are not molecules. Instead, the energy from the photon excites a molecule called 11-*cis*-retinal, which sits in close proximity to the opsin (figure 11).

The 11-*cis*-retinal molecule is made from vitamin A (retinoic acid) and contains a "kink" in its shape that allows it to fit perfectly into a pocket in the opsin protein. Opsins reside in the membranes of rods and cones; the opsin in rods is known as rhodopsin. Only when the retinal molecule and opsin are joined together can they sense light energy.

When a photon strikes a retinal molecule, it imparts enough energy to it so that the kink straightens out. The retinal molecule no longer fits perfectly into the opsin pocket, and it is ejected. Once this happens, the opsin changes its three-dimensional shape as if released from a taut spring. (See figure 11, Light Hitting Rhodopsin, on page 117.) In its new shape, the opsin can activate other membrane proteins in the rods or cones. A quick chain of biochemical events follows that ultimately causes electrical signals (movements of ions) to travel through the optic nerves and into the back of the brain, a region known as the occipital lobe. It is in the occipital lobe that the brain interprets what was seen.

In human beings, there are only four opsins, three of which can sense color. The opsin in rhodopsin senses very low-level light, such as occurs on a starry night, but it does not sense color. The color-sensitive opsins, on the other hand, cannot sense low-intensity light and are mostly useful in daylight. This is why very dim stars generally appear white, because their low-level light is only detectable by the color-blind rhodopsin molecules.

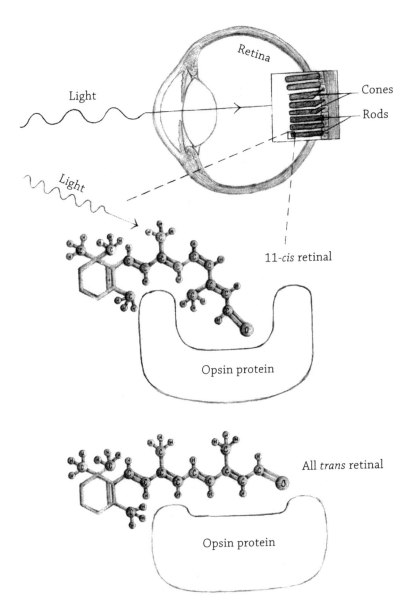

Figure 11. LIGHT HITTING RHODOPSIN
Photons striking a retinal molecule cause the molecule's shape to change, a process called photoisomerism. The retinal molecule is then released from its tight association with an opsin protein.

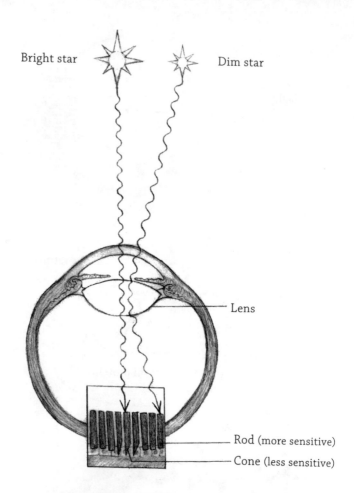

Bright star

Dim star

Lens

Rod (more sensitive)

Cone (less sensitive)

Figure 12. LIGHT ON RODS AND CONES
At twilight, very dim stars are best seen when their photons are allowed to strike the rod-rich sides of the retina. This is most easily accomplished by looking slightly to the left or right of the dim star, allowing its light rays to be bent by the curved regions of the lens.

A curiosity of the human eye is that the sensitive but color-blind rods are not located in the center of the retina, but rather along its sides. We often fail to notice a dim star at twilight when we look directly at it, but suddenly see it when we move our eyes slightly to one side. Under those conditions, light hits the sides of the retina, where our low-level light detectors reside (figure 12).

The cones contain the color-sensitive opsins and are densely packed in the middle of the retina. Thus, when we look directly at something, our color vision is crystal clear. Defects in the amino acid sequence of one or more cone opsins leads to varying degrees of color blindness. This disorder is more common in males, because the genes for these opsins are on the X, or female, chromosome, and you must have two faulty genes (one on each X chromosome) to be a color-blind female. Since males only have one X chromosome (and one Y), a single defective chromosome is sufficient to cause the problem. For similar genetic reasons, color blindness often skips a generation, from maternal grandfather to grandson.

AFTERWORD

It's hard to accept or imagine that we are but an intricately linked web of electrons bridging one atom to another. Countless bags of protons and neutrons, each surrounded by shells of infinitesimally, unimaginably small electrons, buzzing about at mind-numbing speeds. Every inspiration, every thought, hope, or dream that has ever been dreamed by man or woman since time immemorial, has been the result of a few atoms and molecules fleetingly joining together, possibly opening a channel in a nerve cell membrane through which electrically charged atoms flowed.

But the word *atom*—from the Greek *atomos*, indivisible—is actually a misnomer. There exists another stratum of particles below the atomic level, a world of strange, mysterious bits of matter that make up the very elements of the atom itself. The oddly named quarks are the smallest confirmed bits of matter in an atomic nucleus; three of them are required to combine to form a single proton. Quarks cannot be said to be made up of anything; they are made up of themselves. What would they look and feel like if enlarged to the size of a baseball? They may be the final, ultimate level of matter. They may be the stuff that makes up the stuff of life.

When we die, the molecules in our body do not die with us. They live on in microorganisms, escape to the atmosphere, or get absorbed by growing plants, which are themselves later eaten by a cow, sheep, or maybe a caterpillar. Many molecules are degraded into their component atoms, but the atoms remain intact for eternity. Virtually every atom of every living thing that has ever inhabited the earth is still in existence, somewhere, including within your own body at this very moment. The atoms from the first algae, the first trilobites, the first hominids, the atoms in Cleopatra's heart, in Newton's brain, and in a bit of fungus you may have unknowingly crushed underfoot yesterday—all of them remain "alive" to this day. You may even have a piece of Caesar in you—we are immortal.

And so is every other living thing. If you eat meat, you are composed—molecularly speaking—of many different pigs, cows, birds, and fish. And even if you don't eat meat, you have bits of animals inside you, since the nutrients and nitrogen that feed—and become part of—the plants you eat come in part from decaying flesh and animal waste. The atoms and molecules that are in your body right now depend on what you've eaten recently. Next week, your molecular makeup will be considerably different from what it is today, as new molecules are ingested and incorporated into the body and others are eliminated or reconfigured into new molecules. We— plants, animals, protists, fungi, and people—all share a communal pool of atoms and molecules. And, eventually, we all return to that pool what we have taken.

ACKNOWLEDGMENTS

I thank John Michel of Morhaim Literary Agency, not only for his editorial efforts but for the genesis of this book's concept. Thanks to John Parsley, Times Books, Henry Holt and Company, for directing an expert critical eye to the manuscript. I'm grateful as well to Dr. Hershel Raff (Medical College of Wisconsin) and Drs. Charles K. Levy and Tom Gilmore (both of Boston University), who expertly proofread portions of this text. Any errors that may remain are entirely the fault of the author. I thank the National Science Foundation and the National Institutes of Health for providing me with the means to continue my research into the formation, regulation, and actions of those molecules particularly dear to my heart, the hormones. Finally, I am grateful to

Boston University for providing me with the facilities and resources needed to conduct that research.

And, as always, thanks to Maria Widmaier, my first-pass editor and my superior in crossword puzzles and all things literary.

INDEX

ABOUT THE AUTHOR

ERIC P. WIDMAIER, PH.D., is the author of *Why Geese Don't Get Obese (and We Do)*. Dr. Widmaier's major areas of research interests are the control of hormones, regulation of body weight and metabolism, and neonatal development of the adrenal glands. He is a professor of biology at Boston University and has written numerous articles for scientific and nonscientific publications, as well as a textbook on human physiology due out in 2003. He lives in Massachusetts with his wife, Maria, and children, Ricky and Carrie.